FIREFIGHTING DURING WOR[illegible]

Paul Ditzel

Fire Buff House Publishers
New Albany, Indiana 47151-0711

Library of Congress Cataloging in Publication Data
Ditzel, Paul

Firefighting During World War II
Library of Congress Catalog Number: 94-061280
ISBN 0-925165-16-6

FBH Publishers, P.O. Box 711, New Albany, IN 47151-0711

Printed in the United States of America

Typography and Layout: Pam Jones
Cover: Ron Grunder

On the cover:
London Blitz firefighters battle incendiary bomb-caused flames consuming buildings along this Central London street.

To My Loving Wife

Rita Gross Ditzel

Everlasting Friend,

Soulmate and Colleague

Selected Bibliography of Paul Ditzel Books

A Century of Service —
The History of the Los Angeles Fire Department

Emergency Ambulance

Fire Alarm! The Fascinating Story Behind
The Red Box On The Corner

Fire Engines, Firefighters

Fireboats, A Complete History of the
Development of Fireboats in America

How They Built Our National Monuments

Railroad Yard

The Complete Book of Fire Engines

The Kartini Affair

True Blooded Yankee

Quantrill - The Civil War's Wildest Killer

Admiral John Winslow - Naval Hero of the Civil War

FOREWORD

Although I did not realize it at the time, this book germinated in 1945 while I was a United States Marine Corps infantryman, part of a massive invasion force about to storm the beaches of the Japanese mainland. I always knew I wanted to write and, if I survived the invasion, hoped to find work as a newspaper reporter.

Hiroshima and Nagasaki undoubtedly took me out of harm's way. At the end of the war I became a foreign correspondent for a major New York newspaper. I met the late Chief of the London Brigade, Frederick W. Delve. Justifiably proud of his "Heroes With Grimy Faces," he escorted me around the London Fire Brigade stations including the then still secret alarm office where firefighting efforts were coordinated during the London Blitz.

I regret I did not meet Fire Officer Cyril Demarne, who was awarded an Order of the British Empire citation for his London Blitz firefighting efforts. Demarne's book, *The London Blitz, A Fireman's Tale*, is an excellent account of the Blitz as seen through his eyes and those of his comrades.

While in England I met the editor of a British fire journal who tipped me about "The Best-Kept Secret of World War II." His publication was restricted to a single paragraph mention of that wartime disaster in Bombay. I spent months researching the Bombay blasts and saw my manuscript published as my first book in the June, 1959, issue of *True* Magazine.

My foreign correspondent responsibilities took me to Hamburg where I saw the incredible devastation caused by Allied bombers. I experienced the resentment of Hamburg's people when they learned I was an American newsman. In Cologne I saw the bombed out shells of an enormous cathedral: a profound testimonial to the ungodliness of war.

After my overseas experiences I became a Chicago newsman where I met Fire Marshal Anthony J. Mullaney. I will never forget, as he directed large Chicago fires, his leather-lunged bellowing as he stirred firemen to greater efforts. I had forgotten, however, until I began this book, that Chief Mullaney told me of his role in assessing German fire departments and damage to German cities.

I recently found in my library an out-of-print book, *Fire And The Air War*, edited by Chief Engineer Horatio Bond of the National Fire Protection Association, published in 1946 by the NFPA. The book ideally complements Demarne's chronicle and reprints papers presented by Mullaney and other authorities during the 1946 NFPA Conference in Boston.

Lastly, my profound thanks to my recently-wedded wife, Rita Gross Ditzel, who translated my two-fingered typewritten scratchings into computerized manuscript and captions. Only a saint could have withstood my impatience when writing went slow and hurdles came fast, notably 10 Northridge Earthquake aftershocks in one day.

Nor can I overlook my gratitude to Pam Jones whose picture and page layouts were always designed to please me and intrigue readers. And, finally, my thanks to my publisher, W. Fred Conway, for his patience, understanding and assistance as *Firefighting During World War II* came together and into your hands.

CONTENTS

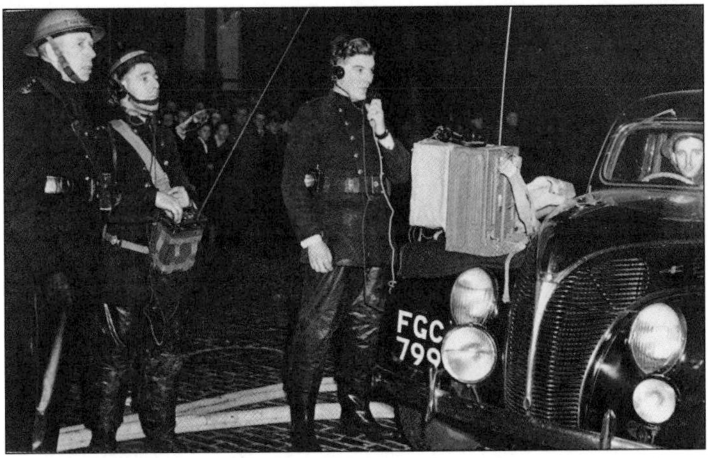

Fire officials direct operations at a major fire during the London Blitz. Mobile command posts were the first of their kind. The idea was later adopted by other fire departments.

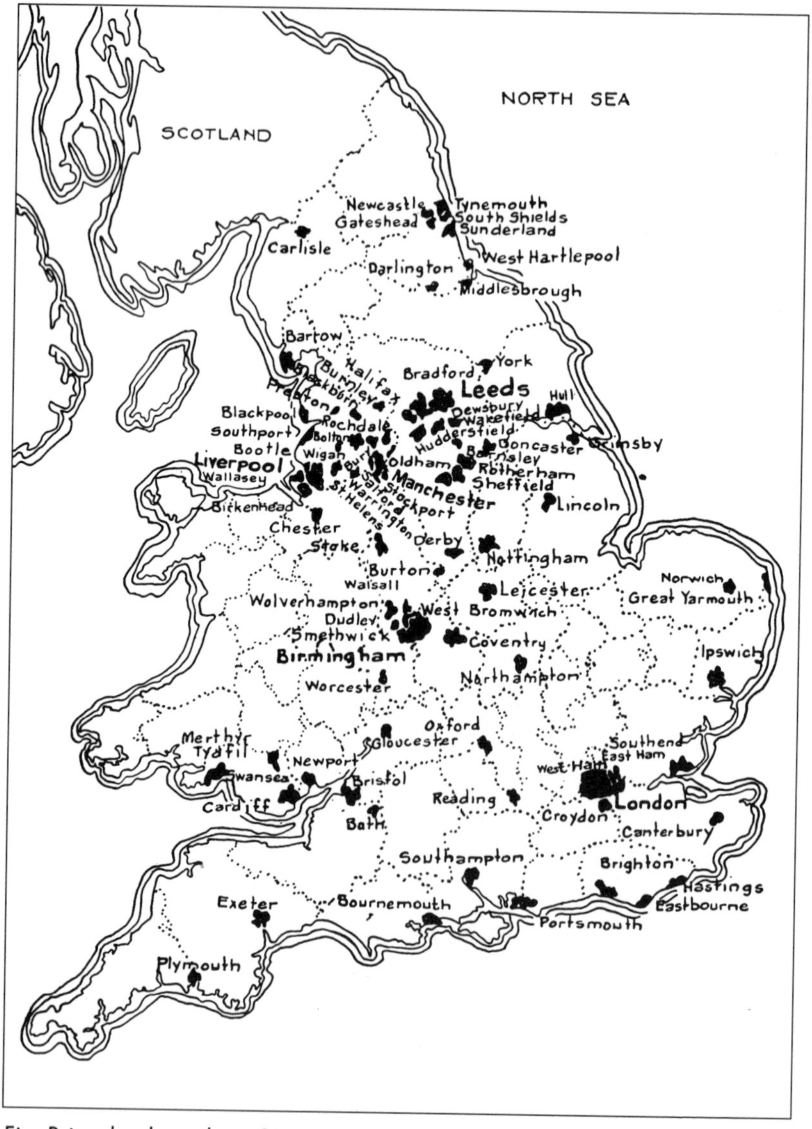

Fire Brigades throughout Great Britain were hard-put to control blazes which damaged about 80 per cent of all buildings. Among the hardest hit were London (lower right), Coventry (center) and Liverpool on the west coast. More than 1000 firemen and firewomen were killed.

CHAPTER I

Heroes With Grimy Faces

Saturday, September 7, 1940, dawned bright and clear in wartime London. During the past weeks German bombers had attacked targets around the city, including the Thameshaven and Shellhaven oil refineries at the mouth of the River Thames.

A Mobile Column of 50 pumpers from the London Fire Brigade answered the mutual aid assistance call to the blazing storage tanks where firefighters were still fighting fires while being strafed by Messerschmitts.

Everybody knew it was but a matter of time before Hitler's Luftwaffe zeroed in on London itself, the world's second busiest port as well the economic, political and religious seat of the British Empire. Both sides of the Thames were crammed with ships, sheds and warehouses stuffed with war materials, including those planned for the invasion of Europe. Multi-story manufacturing and other structures lined narrow streets along and beyond the waterfront.

Among the most vulnerable targets were the three Royal Docks: Queen Victoria, King George V and Royal Albert. Their dockage facilities were among the world's largest and could handle about 50 vessels at a time. East of the docks was the Ford Motor Company plant and a mammoth natural gas works, Europe's largest.

Mobile Columns were self-contained fire departments including 50 pumpers that were quickly sent throughout England to fight fires caused by air raids.

Surrey's Commercial Docks sprawled along the opposite bank of the Thames. Ships were offloading their cargos into hundreds of acres of lumber yards laced by narrow streets made of wood blocks.

Strung for more than 10 miles along and beyond the busy Thames were other conflagration breeders: flour mills, grain storage silos, the Woolwich Arsenal, tar works, alcohol distilleries and chemical plants: all teeming with thousands of workers that balmy September Saturday.

Not far to the west of the docks lay the Houses of Parliament, seat of the British Empire, Buckingham Palace, St. Paul's Cathedral, Westminster Abbey, business and shopping districts as well as homes of millions of Londoners, all oblivious to the final briefings going on at that moment at enemy airfields across the English Channel.

As the day wore on there was increasing hope that London would be spared attack for another day. Since the outbreak of war, September 3, 1939, air raid sirens had been triggered many times for alarms which turned out to be false. There were fears that citizens were becoming overly complacent and would not rush to shelters before air attacks started.

All of England had prepared for bombings. The 1440 fire brigades in the country had recruited additional firemen, formed an Auxiliary Fire Service and trained neighborhood air raid wardens.

Within three years the ranks of firemen would grow to 273,000, plus 70,000 full or parttime women. They worked 48-hour shifts, followed by 24 hours off-duty. When not fighting fires they worked in firehouses producing war materials. Off-duty they held jobs in factories producing more war goods.

During intervals between raids, on-duty London firemen and firewomen assembled war materials. Off-duty they worked in plants producing war materials.

Additional fire stations and apparatus were located in such high-risk areas as the London docks. A top secret alarm center opened near Parliament. A backup system also was in place.

Some of the world's most unique fire apparatus was built. Only time would tell whether these rigs, together with innovative water supplies and communications systems would prove to be effective.

Plymouth made good use of an amphibious fireboat, perhaps the first in the world, to fight wartime tires. The boat saved time by eliminating 30-mile runs along winding streets and over bridges.

Shortly before 5 p.m. that Saturday, the Headquarters Alarm Center logged a Yellow Alert Signal: Coastal and other radar stations were reporting blips showing enemy planes approaching the southeast coast of the English Channel. A few minutes later Yellow Alerts blinked Purple: Hordes of enemy planes were reaching Britain's shores.

Suddenly flashed that most dreaded of all signals: Code Red. Enemy planes were zeroing in on the city. The London Blitz was about to start. Air raid sirens began their moaning wails. Anti-aircraft gun batteries stood ready. Londoners rushed to shelters, while many others, in disbelief and out of curiosity, scanned the eastern skies.

Like swarms of ants, hundreds of high-flying Dornier and Heinkel bombers, escorted by Messerschmitt fighter planes, droned toward London. Wave upon wave of them seemed never ending. The largest mass headed for London's waterfront while others peeled off toward the Surrey Commercial Docks. The staccato chant of anti-aircraft guns pierced the sky with thousands of shells which blossomed into black and charcoal grey puffs of smoke.

First to go were the Ford Motor plant and the gas works which erupted into a never-before-seen fireball over East London. The Royal Docks, sheds and warehouses and acres of tinder dry lumber stacked in the Surrey Commercial Docks were targeted next.

Great screaming sticks of high-explosive and incendiary bombs rained down upon London. Sheds, warehouses, ships, lumber yards, factories and distilleries were quickly set afire on both sides of the Thames. Massive clouds of smoke churned into the late afternoon sky as fires sewn by the bombs took root, sprouted and reached out to spread others carpeting the waterfront.

> "Send all the bloody pumps you've got ... The whole bloody world's on fire!"

Racing toward the flames and smoke shrouding the docks, Station Officer Gerry Knight, 44, his firefighters and auxiliaries had expected major fires, but nothing like these. Knight sent word to the alarm office: "Send all the bloody pumps you've got ... The whole bloody world's on fire!"

The alarm communications network coordinated the dispatch of hundreds of fire engines to Surrey and elsewhere along the waterfront and beyond, while other fire departments were alerted throughout Britain and prepared to hurry to London's aid.

Still the incessant waves of attackers came, dropped their bombs and returned to the continent for more loads. High-explosives snapped water mains as if they

London Bridge remained standing despite incendiary bombings of waterfront buildings and a gigantic lumber yard in the Surrey Commercial Docks.

German bombers target the Royal Docks, among the world's largest, starting on September 7, 1940, the first day of the London Blitz. Grain elevators and flour mills, lower left, are ablaze.

were dry pasta and blasted deep craters throughout the docks. Power, telephone, gas and fire alarm box systems were knocked out.

Hundreds of dull thuds shook the earth as still more high-explosive attacks burrowed home and sent geysers of fire, debris, dust and smoke high into the sky. Brick walls of four and five-story structures — factories, warehouses, apartment buildings — crashed down and blocked London's narrow streets.

Many dockworkers and Londoners who had waited too long to seek shelter were killed. Bodies of men, women and children splattered against buildings or were quickly buried under piles of brick and concrete slabs.

Fire Officer Cyril Demarne recalled: "Columns of fire pumps (500 of them ordered to West Ham alone) sped eastward to attend fires in ships and warehouses, sugar refineries, soap works, tar distilleries, chemical works, timber stacks, paint and varnish works, the humble little homes of the workers and hundreds of other fires that, in peacetime, would have made headline news.

"Two hundred and fifty acres of tall timber stacks blazed out of control in the Surrey Commercial Docks. The rum quay buildings in the West India Docks, alight from end to end, gushed blazing spirits from their doors. An army of rats ran from a Silvertown soap works ... Molten pitch from a stricken tar distillery flooded the road, bringing to a halt all emergency vehicles.

"On the riverside, blazing barges threatened wharves. Some were set adrift by well-intentioned people, only to be swept by the tide downstream, broadside on,

Lumber being offloaded from ships in Surrey Commercial Docks burns furiously during the first raids on that September Saturday. Some 1000 fire engines quickly answered calls to Surrey, across the Thames River from London. Hundreds of waterfront fires in London (background) go unchecked.

Preparing for wartime incendiary raids, London bought 20 fireboats to buttress its existing fleet of five.

to the peril of fireboats maneuvering to pump water ashore to feed the land pumps."

Pre-planning for the blitz proved to be effective, but none of it could foresee the effects of saturation bombing. Before the Blitz, 20 new fireboats were built. These boats had twin 1000 gallons per-minute-pumps mounted amidship to supply land companies and a large bore stationary nozzle in the bow for fighting ship and dockside fires.

A maelstrom of smoke and flames shrouded the Royal Docks where pre-planning again mitigated incredible firefighting problems. Because bombing would likely prevent fireboats and land-based apparatus from entering the docks, four firefighting barges, *Fred, Moonsoon, Brampton* and *Sweden* were each outfitted with four Dennis fire pumps. Twin holes were cut in the stern of the boats. Hose from the Dennis pumps was fed through the holes and draped over the side. The barges were propelled fore and aft, starboard and port, by the thrust of the water.

Firemen with the two other pumps fought fires aboard vessels and along the wharves.

The 20 boats and four barges had joined the London Fire Brigade fleet, including its flagship, *Massey Shaw*, which had distinguished itself that spring. The *Shaw* was the only fireboat in what Prime Minister Winston Churchill called The Mosquito Fleet which evacuated thousands of English and French soldiers from the shores of Dunkirk. The *Massey Shaw* and its 13-member volunteer crew of London firemen and auxiliaries rescued, while under German bombardment and strafing, more than 600 soldiers and sailors.

Londoners line the banks of the Thames to welcome home the *Massey Shaw*. The fireboat, queen of the London Fire Brigade's fleet, was used by firemen and their auxiliaries, to evacuate more than 600 troops trapped on the French beaches at Dunkirk.

Pump-escapes were a mainstay of Britain's fire departments. Rear-mounted wooden ladders were wheeled into alleys and other tight spots to fight fires and make rescues. Hose was stored in a compartment behind the pumps. These and other apparatus were painted drab grey for camouflage. Black paint on headlights further disguised them from the German Luftwaffe.

Mainstay of London's apparatus fleet were fire pumps and pump-escapes which were engines with a two-wheeled portable wooden ladder mounted at the rear. Pivoting off the rigs, these escape ladders were wheeled into narrow places and raised by ropes and pulleys. Rear-mounted turntable ladders enabled firemen to reach upper stories of buildings or, with a nozzle mounted at the end, could deliver a stream into the flames.

Alongside this frontline equipment were Dennis trailer pumps, mostly manned by auxiliaries. A taxi driver came up with the idea of a quick-hookup device which enabled jitneys, automobiles and light vans to pull them to fires while auxiliary firemen rode inside. This was probably the first time firemen went to fires in taxis, but they had no time to remark about it, what with more fires than firefighters or apparatus.

All this equipment and armies of firemen would have had to stand by helplessly, except for pre-war innovations in water supplies. London taps the Thames which feeds canals and underground mains. Figuring, correctly, that traditional mains would be destroyed by air raids, 6-inch steel piping was laid in gutters along streets. Hose connections spaced along the system could each supply five streams.

Another bit of pre-war ingenuity were large lorries which resembled vans. Pre-connected hose, 2 1/2, 2 3/4, 3 1/2 and 7-inch in size, was folded from floor to ceiling. Some 6000 feet of this rubberized canvas hose could be laid when the rear door was raised to permit the hose to spill out as the lorry moved forward.

The 7-inch supply used for long relays, provided the capability of delivering 1600 gallons-per-minute at 50 pounds pressure per-square-inch.

Thousands of camouflaged water tanks of from 3000 to one million gallons capacity were built in open areas and along London streets. High-explosive hits that afternoon provided somewhat of a surprise: While fixed underground systems shattered, the flexible street piping seemed better able to withstand bombs.

Backstopping even this system were tractor-drawn flatbed trailer hose lorries (wagons), each carrying 1000 feet of steel pipes. Nine firemen, using iron tongs, laid and coupled 1000 feet of water pipe in 16 minutes as buildings blazed and collapsed around them.

"Under good conditions, 19 men could lay one mile of 6-inch piping weighing 40 tons and have water passing through it in 45 minutes," said Commander Sir Aylmer Firebrace. Forty stationary pumps, moreover, were installed along London bridges. Suctioning from the Thames, they were each able to supply up to 3000 gallons per minute.

As more and more squadrons of bombers sewed more fires late that afternoon, firemen braved explo-

London firemen use tongs to couple lengths of steel pipe for water supplies. While fires raged around them they connected 1000 feet of pipe in 16 minutes and quickly had hoselines operating.

Lorries resembling moving vans carried more than a mile of pre-connected hose which spilled in a continuous line to provide more backup for besieged firemen.

sions while lugging their hose over debris in attempts to surround fires which were threatening to become conflagrations. Intense heat seared their unprotected faces. Back pack respirators were of little value, but helped some amid the toxic gases lurking in the smoke.

By now some 500 pumps were battling burning lumber stacks aboard ships and in the Surrey Commercial Dock yards. Five hundred more would be there before dawn as firemen and apparatus from Birmingham and other areas raced to help the London Fire Brigade. Water turned to steam before it penetrated the blazing lumber. Much of it ran down the wood block streets. Firemen sensed the ground rising and falling and realized the blocks were being forced upward by the runoff.

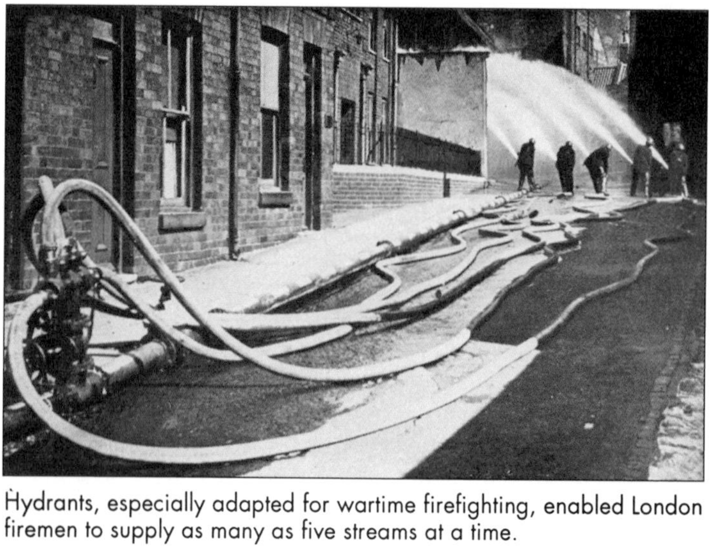

Hydrants, especially adapted for wartime firefighting, enabled London firemen to supply as many as five streams at a time.

Firemen worked for two hours to free this London housewife trapped in the rubble of her home. She survived, but hundreds of others buried inside their bombed buildings did not.

At 6 p.m. the droning of enemy aircraft grew silent and sirens sounded the All Clear signal. Only an hour had passed since the attacks began, but those minutes seemed like a millennium. Londoners came from their shelters and were stunned by the devastation, the blankets of smoke and flame and the sight of burned, blackened and dismembered men, women and children; knowing all the while that many more were buried alive and dead under the rubble. Firemen, auxiliaries and civil defense workers, their faces grimy from smoke, sweat and dust, started search and rescue work.

Bomb and fire-destroyed walls of buildings crashed by the hundreds into London streets. Debris-clogged streets were quickly cleared to provide continuing firefighting access. Open streets were a morale-booster for Londoners.

Vast areas of the waterfront and beyond also were burning beyond control. The four firefighting barges in the Royal Docks were fighting ship, dock and warehouse fires. Further from the waterfront, turntable ladders somehow managed to take up positions along streets as tall buildings blazed on all sides. Their streams could but delay the spreading flames from block to block, street to street, neighborhood to neighborhood until they melded into more than a dozen conflagration-size fires. The only good news was that Hitler had spared St. Paul's, Parliament, Buckingham Palace, Westminster Abbey and other historic landmarks; some of them dating to medieval times.

Alarms were flooding into dispatching consoles from areas where telephone and fire alarm box systems still worked. As large blazes melded into giant fires, the women dispatchers sent self-contained fire departments—Mobile Columns of 50 pumps—similar to the one which was ordered to the still-burning Thameshaven and Shellhaven refineries outside London.

Two hours later the Luftwaffe returned and renewed its blitz, concentrating upon the waterfront and the manufacturing, commercial and residential districts immediately beyond. Tracers from anti-aircraft shells probed the darkness as waves of planes blasted London with high-explosive bombs and incendiaries. Hundreds more fires were erupting where none had been minutes earlier.

Firemen vainly tried to save this commercial building during a nighttime raid in the Piccadilly section of London.

Among the targets was a paint factory which took a direct hit and was fully-engulfed when firemen arrived. The officer-in-command started to call for more men and apparatus. To their astonishment another high-explosive plopped inside the plant and exploded with such force that it blew the fire out.

The onslaught continued throughout the night as entire neighborhoods of buildings were set afire. Bombs ripped apart railroad yards. Fifty-foot lengths of track speared apartment and other buildings. Shell fragments—both from enemy and friendly fire struck sparks off firemens' metal helmets as the earth trembled with thunderous explosions.

Tracks from blitzed railway yards vaulted buildings while slamming into others in a tenement area beyond the waterfront.

London's Royal Docks, one of the world's largest, was sledgehammered again and again by bombers. Rebuilding was not completed until years later.

As many as 3000 fire alarms during raids were received at this London Fire Brigade Control Room. Operators coordinated appropriate responses of firefighting personnel and apparatus.

London firewomen carried messages between firegrounds and alarm centers when raids knocked out communications systems. Speeding around burning buildings, toppled walls and bomb craters was especially hazardous for the motorcycle riders.

Alarm headquarters was deep inside a building near the Thames. Its fort-like steel and concrete walls and 12-foot-thick ceiling were built to resist bombs, fire, smoke and gases. Firewomen worked by candlelight while coordinating deployment of fire apparatus and firefighters.

Although the main communications center was equipped with two-way radio, wartime priorities precluded installation of this equipment on most pieces of fire apparatus. Communication was mainly by telephone and thousands of street corner fire alarm boxes, especially in waterfront districts where many languages were spoken. A numerically coded firebox, when its lever was pulled to transmit an alarm, spoke in a universal language.

As feared, many telephone and firebox systems were quickly destroyed in areas where they were most needed. Firewomen, trained as motorcycle couriers, raced back and forth from firegrounds and communication centers. Still more messengers had been recruited among youths who were at least 16. Many lied about their ages and were signed up anyway.

All through the night these couriers carried messages; their headlamps painted black except for tiny slits which were not much help in debris-cluttered streets. Steering around rubble and bomb craters, dodging falling walls and being thrown off their bikes by explosions, they nevertheless provided effective communications backup while managing to avoid unexploded bombs littering streets as a vast canopy of flame and glare etched London's skyline.

Her Majesty the Queen and her royal entourage inspect London firewomen.

Thousands fleeing from adjacent residential districts collided with apparatus and firefighters dragging and shoving their Dennis trailer pumps to fires while others strung out hoselines they lugged on their shoulders.

As firemen and women were joined by those from Birmingham and other British cities, all-night battles continued, street-by-street. It was impossible to surround fires because of the heat and smoke blown into their faces. Firefighters established battlelines and made determined stands which they hoped they would not be forced to abandon. Thundering crashes of brick walls of three, four, five story and taller buildings toppled into streets and alleys.

Around dawn, with 1000 pumps suctioning water from the Thames near the Surrey docks and similar battlegrounds nearby, the German planes called it a day and returned to the continent. As fast as they left, firefighters attacked rubble with bulldozers to open streets for firefighting, search and rescue operations, while anticipating the momentary resumption of bombing.

By dawn there were 19 conflagrations in London, plus 17 major fires requiring at least 30 pumps and firemen, in addition to hundreds of smaller blazes. Thousands more were doused by civil defense workers. Uncounted numbers of Londoners had been killed or injured.

Hundreds were homeless and had lost all their belongings. Especially hard hit were residential areas of waterfront workers, their families, relatives, neighbors and friends. Rescuers and others digging through debris were forever traumatized when they came upon bits and pieces of bodies.

London Fire Brigade women auxiliaries set up mobile canteens along firelines and served tea to firefighters looking skyward while awaiting resumption of German air raids.

The Luftwaffe did not return that day but it was back again that night and, for eight more hours, bludgeoned London. Among the fire companies racing to fight fires along Abbey Road were those under Station Sub-Officer Walter Turley and his men. They found a building well-involved and laid a hoseline through the front door. Without warning the building collapsed, killing 11 firemen:

Turley, two auxiliary firemen and eight civil defense corpsmen. The dangers of firefighting know no age limit or gender. The 11 dead were the first of hundreds of men and women firemen of all ages who would be killed during the London Blitz.

Royal Air Force Hurricane and Spitfighter planes swarmed aloft and swooped down upon the raiders. In less than a week they drove the Dorniers, Heinkels and Messerschmitts from the daytime skies over Britain. Never again would the Luftwaffe be able to mount significant daytime attacks. But for 57 consecutive nights the Germans continued raids and only then took a short break because of bad weather over the continent. After that the night after night onslaughts continued.

Power, water, gas and telephone systems were quickly restored, only to be blitzed again and again and just as quickly put back in service. Bomb craters along building-demolished streets were filled with water and served as reservoirs for firefighters. They battled flames during the day so they could report at dusk: "No fires showing a light after blackout time." But the Thames often betrayed them as moonlight caused phosphorescent-like glows and provided enemy planes a surefire route into London. No fire, however trifling, was left unattended as British firefighters snuffed whatever could guide the air invaders.

Eleven London firemen were killed when the floors of this building collapsed. They were the first of hundreds of firemen who would be killed fighting Blitz fires.

Munitions depots, including London's Woolwich Arsenal, were frequent targets of Dornier and Heinkel bombers. The crew of a fire station near Woolwich headed in that direction whenever air raid sirens warned of incoming waves of Luftwaffe planes.

The Woolwich Arsenal was an almost nightly target. Explosions and fires rocketed ammunition high into the night sky, burst into firework displays and rained shrapnel upon firefighters.

A high-explosive bomb scored a direct hit on a gigantic flour and grain complex, sending a fireball boiling high into the air.

Starting a month after the Blitz began, bombers were still attacking the Royal Docks where barge firefighters performed heroic deeds despite thick clouds of acrid smoke and searing flames. Ships and docks and warehouses burned beyond immediate control. Some enemy squadrons now turned to carpet bombing of the square-mile that is London City itself and surrounding commercial and business districts and residential areas. St. Paul's Cathedral was struck on October 1, by the first of an eventual 800 bombs and would thereafter be a frequent target.

Shortly before 10 o'clock on the Blitz Night of Sunday, December 8, Station Officer Fred Dell and his firefighters in No. 16 Station in the West Ham District were preparing to respond to an alarm when a direct hit destroyed the firehouse. Dell and his nine fellow firemen were killed. Services for the dead were conducted by Vicar W. D. Tarley of St. Matthias Church. Two weeks later Reverend Tarley was killed during an air raid.

A month after Dell and his firefighters were killed, enemy raiders blasted Coventry in central England for 11 consecutive hours. The Cathedral of St. Michael,

Londoners pay their final respects to seven firemen who were killed when their firehouse took a direct hit on December 8, 1940. Union Jack flags are draped over their coffins aboard pumpers which they had used to fight fires.

built in the 14th Century, was destroyed. Mindless attacks on this and other landmarks became known as Baedekers, after the guide books of that name.

Other British ports and cities came under the German bombardiers' sights: Liverpool, Portsmouth, Plymouth and Manchester. Bombs struck an oil tanker off Liverpool. Gobs of black smoke and flames erupted from the vessel's storage compartments as the crew abandoned ship. Firefighters set out in motorboats with foam, controlled the fire and brought the ship and crew into Liverpool.

Butterfly bombs snagged rooftops and exploded by vibrations.

The London Blitz continued night after night, each raid as vicious, if not more so, than the previous night. When it seemed that London could take no more pounding, the Heinkels and Dorniers returned en masse just after nightfall, Sunday, December 29. This time the target was the square mile heart of London.

In this tightly compacted area five miles west of the Royal Docks was St. Paul's Cathedral, whose religious and historical significance went back to the early 7th Century. It had withstood the ravages of invaders as well as The Great Fire of London in 1666.

In the immediate vicinity were block upon block of four and five-story brick and masonry structures, many of them hundreds of years old: churches, libraries, the nub of Great Britain's book publishing industry. Streets, as they were in medieval times, were intertwined with even narrower alleyways. None of the buildings were protected by sprinkler firefighting systems.

Even if they were, the sprinklers would have been overwhelmed by high-explosive bombs and the virtually total lack of firestopping walls. London firefighters had never experienced so many hundreds of incendiary bomb clusters which seeded fires in so concentrated an area. Buildings on both sides of the streets for blocks around quickly became engulfed.

While fighters connected lengths of steel water supply pipes from lorries, other trucks laid supply hoses to the waterfront. Fireboats tied up along the bank of the muddy flats. Other fire engines were spotted at 700-foot intervals to maintain pressure along the half mile from the Thames to firegrounds. The specially-equipped hydrants along Blackfriars Bridge were also tapped.

Individual fires joined others along streets and became as one. Tornadoes of flame sucked in wind from street level and broadcast flaming embers into the faces of firemen. Above the din of churning fire pumps could be heard screams of firemen, firewomen and teenaged youths as the embers wormed inside their helmets and protective flaps to sear eyes, ears and raise blisters on their necks. A fireman was struck by an incendiary

Six-inch steel piping provided backup water supplies. Nine firemen laid 1000 feet of pipe in 16 minutes.

A fire zone was declared and communications were coordinated from mobile command posts as occurred during this aerial view from St. Paul's Cathedral.

bomb which impaled itself in his back. Screaming, he fell to the street as the intensely burning white phosphorous and its molten metal burrowed into him until he fell silent.

Firefighters dragged their Dennis pumps over rubble and broken glass, stumbling all the while as shards tore into their hands and faces. Their feet, despite boots, were rubbed raw as they scrambled.

Firemen under doorways (above) take refuge from falling walls, debris and shrapnel during the height of nightly raids. At dawn, when the raiders left, they moved in (right) to mop up overnight fires.

Whenever they dared, firemen stole glances above them to see if brick walls and ornamental masonry were about to fall. Where turntable ladders were able to take positions, the wind-blown fury of heat, smoke and flying embers quickly forced officers to order the nozzlemen off the ends of ladders.

Toppling walls crushed eight firemen. Eight more would be killed before the night was over. Mounds of bricks, glass and other debris forced firemen to retreat to fighting fires from doorways across the street from burning buildings as clouds of steam rose from the water runoff. They made stands for as long as they could until they were forced backward.

Other firemen found themselves trapped inside the hellish inferno. Slipping, staggering, almost blind from the smoke and heat, firefighters retreated as huge paint blisters swelled on the apparatus. Turntable ladders were destroyed, wooden ladders on pump-escapes burned first before the flames grotesquely twisted their metal sides, hoods and cabs. A group of firefighters finding themselves hemmed on all sides by fire groped their way to a subway grating, peeled it back and found refuge in the tunnel below.

Still the bombers came and still more men and apparatus were sent to the heart of London as 2000 pumpers and 9000 firemen, firewomen and messenger boys worked together. London's literary center — Paternoster Row — was nevertheless consumed by fires.

As flames encroached upon St. Paul's Cathedral from all sides, Prime Minister Churchill sent word that it must be saved at whatever cost. The cathedral was more than an edifice to the British. It was a sign that so long as St. Paul's stood, so would the resolute Britons.

In this, one of the most famous photos of the London Blitz, flames encroach upon St. Paul's Cathedral which firemen saved in late December, 1940.

Prime Minister Churchill sent word that St. Paul's must be saved at all costs.

Nozzles at the ends of these turntable ladders provide effective water curtains as London firemen checkmated flames before they reached St. Paul's Cathedral.

The battle to save St. Paul's raged on into the night. Firefighters did not need the prime minister's call. They made incredibly heroic stands against the flames, smoke, flying embers and crashing walls as well as those threatening to fall at any moment. Pumpers churned. Their exhaust pipes turned red from the engines' throbbing for hours.

In what would go down as one of the supreme firefighting feats of the London Fire Brigade, the flames approached St. Paul's and were stopped.

St. Paul's ordeal was shown on theater screens the world over and the nightly raids on London were described in live radio broadcasts to the United States by Edward R. Murrow from the British Broadcasting Centre in London. The full story of the Blitz would not be told until many years later. Casualty figures and destruction of strategic military facilities were closely guarded secrets by the British Government so as not to give encouragement to the enemy.

This fire engine fell into a bomb crater and was destroyed during a nighttime raid of December 30, 1940.

Flames ravage buildings during the height of one of the worst nighttime raids.

Prime Minister Churchill called all firefighters: "Heroes with grimy faces."

The Luftwaffe needed no encouragement as they returned nightly with clocklike-regularity. Little did Londoners know the worst was yet to come. Starting at 8 pm., Wednesday, March 19, 1941, there began England's worst air attack of the war with raids that touched off 2000 fires. As many more were put out by air raid wardens and their neighbors using hand-held water pumps, sand for smothering burning magnesium and water buckets. Out of these raids came still another unique fire apparatus: Soap Box Fire Engines. Fashioned of cardboard boxes and with baby buggy wheels, these Soap Box rigs carried hand pumps, sand bags and water buckets.

The raid that night would be remembered as the first use of parachute mines. They were as large as gas station storage tanks and fell silently and without warning, unlike the shrieks of bomb clusters.

A fire crew was answering a call to a fire near the Royal Docks when their apparatus and all five men aboard were obliterated by an exploding mine.

With little respite as nighttime raids set fires that were battled into and beyond the next day, Londoners were again challenged starting around 9 p.m., April 16. Seven hundred German bombers unleashed 150,000 incendiaries and 800 tons of high-explosive bombs which set 2250 fires. Starting along the banks of the Thames and stabbing inward, the city and surrounding areas were quickly alight. Silhouettes of a panorama of flames and smoke etching the night skies had become an almost nightly affair.

A high-explosive bomb pierced the roof of St. Paul's, burrowed into the crypt, and exploded to create a large crater. Still St. Paul's stood. Fierce fires racked the

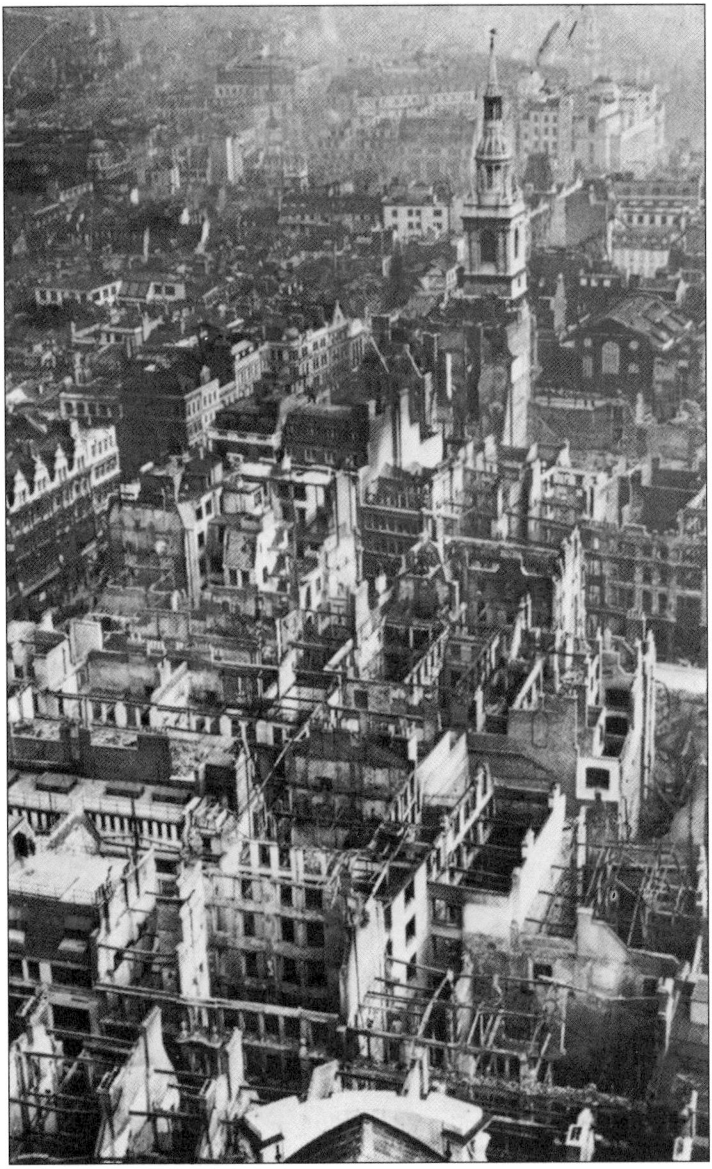

Much of central London lays devastated with only burned out hulks of brick and stone buildings still standing.

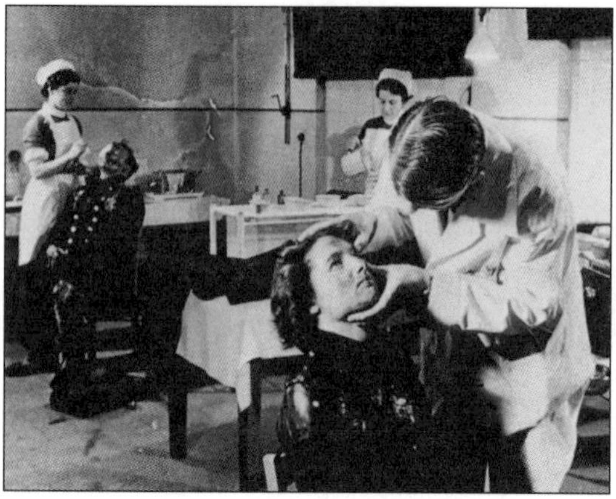

Firemen and firewomen are treated at a London hospital for injuries suffered on the night of April 16, 1941.

commercial and business areas of the central city. A bomb detonated when it hit the British Broadcasting Centre headquarters, injuring 50 broadcasters and technicians. Their rescue was impeded by 50 tons of phonograph records which cascaded onto them.

Near the Royal Docks portable monitors were each delivering 600 gallons per minute from their 1 1/2-inch nozzles.

One of these portable monitors broke loose and impaled a fireman.

Under intense back water pressure, the monitors were held in place by twin steel feet at the front and a footstand at the rear. One of them broke loose and lashed wildly. Fireman A.B. Advani had come from his native India to apprentice with British firemen.

One of the front struts of the flailing portable monitor pierced his thigh and narrowly

The British Union Jack hangs from the window of a bomb-shattered building as the King and Queen reassure Blitz survivors.

missed major arteries. Bleeding, and with the strut still protruding, Advani was rushed to a hospital. He recovered and returned to India where he became New Delhi's Chief Fire Officer.

Others were not so lucky. Twelve hundred Londoners, including six firemen, were killed in that raid and thousands of citizens injured before the Luftwaffe returned home at dawn. London's agony seemed unending. Still they stood steadfast and defiantly draped Union Jacks out of bomb-shattered windows.

The deadliest raid of all started on a clear spring night, eight months after the London Blitz began. A full moon bathed the city in a macabre glow which made targets more easily seen by the Luftwaffe. Even the heavens now seemed to be conspiring against them. On and on roared waves of Dorniers and Heinkels dropping ton after ton of bombs.

Entire business and commercial blocks of tall buildings erupted in flames which quickly lockstepped and overwhelmed other nearby districts. So, too, were firemen overwhelmed. Lack of water made their plights desperate. Hundreds of buildings spouted fires from every window, door and roof. Weakened walls crashed into streets and upon adjacent structures.

The International Headquarters of the Salvation Army was consumed on Queen Victoria Street as nearby buildings also fell victims to bombs, fires and lack of water.

Supply lines were once again laid from the Thames and hydrants along bridges. Hose lorries spilled their beds of hose which flipflopped as it streamed from behind the vans.

A portable reservoir was filled with 5000 gallons of relayed water near one of the worst areas. Firemen dropped suction hoses into the canvas bag and began to supply their engines when a high-explosive bomb slammed smack in the center of their clustered pumpers. Little was found of the 17 firemen who vanished from the face of the earth.

As the number of fires grew to around 3000 and cast their unholy glares over the night skies, more incendiaries and high-explosives splattered down. Parliament's House of Commons was hit. So was Buckingham Palace. And Westminster Abbey, site of coronations and weddings of Great Britain's kings and queens, burial crypts of monarchs, statesmen, national heroes and men of letters, including Robert Browning, Geoffrey Chaucer and Lord Tennyson. Missiles ricocheted off Big Ben.

Firemen hauled their hose and equipment up some 200 stairs and along a maze of the Abbey's catwalks to

Intense heat and smoke forced firemen to retreat before they could get this turntable ladder into the firefight. The Salvation Army's international headquarters (right) was destroyed.

London firemen (top) supplied their pumpers with relayed water which filled these portable swimming pool reservoirs. A bomb hit smack into the middle of the other firemen drawing water from a reservoir (left). Seventeen of them were killed.

battle flames chewing across the open timber ceiling, one of the world's prime examples of medieval architecture. They managed to put out flames but not before parts of the roof collapsed, barring their escape. Firemen tightly secured their hose lines to the tall rafters and slid to safety.

Around 1500 men, women and children were killed that night, including 35 firemen. Thousands were injured. Londoners felt slight consolation in news that the Royal Air Force had shot down two dozen of the attacking planes. They feared that hundreds more would quickly replace them.

It was not to be. That May 10, 1941, raid was the last significant bomber attack on London. Several weeks later the Luftwaffe turned its ferocity on the invasion of Russia. Soon came the V-1 and V-2 rocket attacks on England. They caused heavy casualties and damage, but the main work for firemen, civil defense workers and others was search and rescue in the ruins of rocket-shattered buildings.

Nothing would ever eradicate the memories of 14,000 people killed in Britain during the raids. The more often England, Scotland and Wales had been pounded, the more resolute their people became that they would survive whatever Hitler hurled at them. Thousands of buildings lay in ruins. Damage assessment teams reported that at least 80 percent of all structures in the country were destroyed or damaged.

Firemen battling blazes in the open attic of Westminister Abbey were trapped by falling debris. Tying their hoses to roof supports, they slid to safety on the Abbey's floor.

V-1 and V-2 rockets caused widespread damage. While few fires resulted, rescue operations were especially dangerous because of the threat of imminent collapse of walls.

The horrors of nighttime bombings are mirrored on the faces of London firemen and firewomen who await treatment for their injuries.

As despicable as were the raids against non-military targets, Britons would forever remember, too, the Luftwaffe's failure to break their morale by attacking historic religious, cultural and political centers.

Nor have the 273,000 firemen and 70,000 women who stood steadfast together been forgotten. Altogether, 1027 firemen and women were killed. Their names are engraved upon a Memorial in old Change Court, not far from St. Paul's Cathedral. Prime Minister Winston Churchill called all firefighters: "Heroes with grimy faces."

More than 100 people, mostly housewives, were killed when a rocket bomb struck Smithfield Market. At right, firemen remove the body of a victim from inside the market.

A panorama of smoke, flames and destruction sprawls across London in this photo taken from St. Paul's Cathedral. Firefighters' supremely successful efforts to save the cathedral are regarded as among the most heroic feats in the history of the London Fire Brigade.

St. Paul's Cathedral remains standing despite London Blitz devastation for as far as the eye can see.

CHAPTER II

A Day To Remember

Anthony J. Lopez, an off-duty Honolulu firefighter, was driving to church shortly after 8 o'clock that peaceful Sunday morning, December 7, 1941, when a low-flying airplane barely cleared the treetops. The Rising Sun insignia of Japan was under its wings.

Lopez, later deputy chief of the Honolulu City and County Fire Department, recalls he was not alarmed. "News reports had alerted us to imminent war games at Pearl Harbor and other nearby military bases. So I passed off the Japanese markings as part of the make-believe."

Shortly before church services started, the building was rocked by a distant explosion. Lopez guessed it was part of the realism of the war games.

"About half an hour later, fire engines passed the church. Then I heard more apparatus and knew a second alarm had been pulled. I went outside and saw a column of thick, black smoke several miles to the west. It still did not occur to me that we were being attacked. I guessed that some of the Navy's fuel storage tanks at Pearl Harbor had ignited.

Lopez knew that a large fire would result in a recall of off-duty firefighters, so he hurried home to get his firefighting turnouts.

"I tuned my car radio to a local station for news. A disc jockey said we had been attacked by 'a foreign country.' Then the station went dead. I tuned to the two other Honolulu stations. They were dead, too. Only then did I connect the Rising Sun insignia to the explosion and realized the Japanese were attacking Pearl Harbor.

Lopez' disbelief was mirrored throughout Honolulu, which reacted slowly to the methodical destruction of America's Pacific fleet along Pearl's Battleship Row and Army air power at Hickam Field as well as other U.S. military installations. Lopez refreshed his memory by referring to Honolulu fire records of the 39 alarms answered on Pearl Harbor Day.

Indicative of the slow reaction was the first alarm logged at 8:26, half an hour after the initial attack. It was a telephone call for a second alarm mutual aid response to Hickam Field, the Army's main air base in the Pacific. This was the dispatch which Lopez heard.

Engines, hosewagons and Assistant Chief Edward P. Boyle turned out to Hickam. It was the last alarm for three of the 29 officers and men who answered the call. Arriving at Hickam they found the base firehouse in flames and both rigs destroyed.

Water mains were blasted by bombs. The pumpers drafted from water in a bomb crater. Firefighters laid 6400 feet of hose to attack the flames raging through hangars, barracks and neat rows of planes that were sitting ducks for the attackers. The second wave of Japanese planes came in low, firing machine guns and dropping more demolition bombs. Lopez doubts the victims heard them.

The *U.S.S. Shaw* explodes after her forward magazines ignite. Anthony Lopez mistook the smoke for oil storage tanks. Only after he saw the Rising Sun insignia did he realize the Japanese were attacking Pearl Harbor.

U.S.S. Arizona Memorial
National Park Service

Killed by demolition bomb fragments or machine gun bullets were Fire Captains John Carreira, Thomas S. Macy and Hoseman Harry T.L. Pang. Six other firefighters were wounded. "My friend, Engineer Lawrence Kanoa, saved himself by diving under his rig," said Lopez. The apparatus was badly shot up, but make-do repairs were quickly made.

Lopez was leaving home as his captain, Joseph Pacheco, drove by. Lopez joined him. Their firehouse, Waikiki No. 7 Engine, was empty. The company had been ordered to fill in for Makiki Station 3 and both companies were subsequently ordered to the command post at the Central Station on Beretania Street in the downtown district. That's where Lopez and Pacheco finally caught up with their company.

By then, black smoke was billowing from residential and commercial districts. These fires were thought to have been started by Japanese bombs and saboteurs. It was later determined that the blazes were touched off by faulty or incorrectly timed anti-aircraft shells fired by Americans. The shells peppered the tinder dry wooden frame cottages and tenements dotting the city. When they exploded they slaughtered civilians and touched off fires.

The number of alarms overwhelmed the department, which had a normal strength of 161 officers and men, eight 1000 gallons-per-minute pumpers and hosewagon companies, each of which ran as one outfit, and a 75-foot aerial ladder; all Seagraves. The sprawling rural area of Oahu with its valuable pineapple and sugar cane fields was protected by only two 500 gallons-per-minute Seagrave pumpers.

Fire Chief W.W. (Bill) Blaisdell was confronted with a problem unparalleled in any American city. With

Sailors at the Ford Island seaplane ramp look on as the *U.S.S. Shaw* explodes — the flames at the right are from the U.S.S. Nevada.

U.S.S. Arizona Memorial National Park Service

Fire alarm boxes were being pulled all over Honolulu, with one round coming in on top of another. Chief Blaisdel ordered dispatchers to ignore box alarms and to send apparatus only to addresses given by telephone callers.

the department's strength depleted by three companies that would be committed all day at Hickam, how much apparatus and manpower could he commit to other fires? Would there be enough men and equipment in reserve to handle more fires if the Japanese continued their attacks through the day as everyone expected? Bad as the fires were in Pearl Harbor, there was nothing that could be spared to help the Navy firefighters.

"There were so many fire boxes being pulled that one round was coming in on top of another," says Lopez. Chief Blaisdell ordered dispatchers to ignore all box alarms and respond apparatus only to addresses given by telephone callers. He wisely split his remaining five city companies. Engines and hosewagons would run separately. The pumpers were loaded with extra hose. Hosewagons had to make do by coupling their standard firehose to hydrants nearest the fireground and operating with the pressure in the mains.

Nine commercial trucks were commandeered and loaded with hose. These rigs were manned by a few firefighters, but mostly city and county employees and civilian volunteers. "The rigs carried 500 feet of hose, a nozzle and a hydrant wrench," said Lopez. Fortunately, the department had the foresight to prepare for war and had laid in extra hose, siamese connections, hydrant wrenches and other equipment. "Even with these makeshift rigs, the potential for disaster still remained and we all knew it," said Lopez.

Drawing from official records, Lopez provided a sampling of the alarms soon after the first bombs hit:

8:33 a.m. Shell hits a two-story wooden dwelling, 2683 Pacific Heights Road. $3000 damage.

8:44 .a.m. Shell explodes and shatters the sprinkler system in the 3-story brick-and-wood Lewers & Cooke Department Store in downtown Honolulu. $5760 damage.

9:25 a.m. Enemy plane crash. Depot Road, Waipahu.

9:26 a.m. Shell explodes in a wood cottage at 1630 Leilehua Lane. Edward Jona, 19, killed.

10:50 a.m. Shell penetrates the floating roof of a one-story steel tank of natural gas at the Honolulu Gas Co., Ltd., Prison Road and Pacific Streets.

"This was one of the most spectacular fires of the day," recalls Lopez. "Firefighters attacked with fog nozzles." Ordinarily, that fire would have been good for extra alarms, but Pawaa No. 2 engine battled it alone until the fire was controlled 1 1/2 hours later.

11:55 a.m. Shell hits houses and two-story wood frame tenements and stores at 1935 S. King Street near McCully Street in the district between downtown Honolulu and Waikiki Beach. Three adults and a 3-month-old infant were killed.

"The fire started in a cottage owned by Firefighter 'Tarbell' Kinney," said Lopez. "That fire was our worst of the day." Flames, helped by a stiff breeze, eventually spread throughout the block, destroying 13 buildings. Thirty-one families were left homeless. Most of them were Japanese.

The fire which normally would have brought out nearly all the department and chiefs, was handled by Lieutenant Edward P. Meek and a few hosewagon firefighters using only hydrant streams. The fire communicated from the block of blazing buildings to the wood shingle roof of the nearby Lunalilla School. Somehow the firefighters managed to make a good stop before the fire burned below the second floor.

About this time another Japanese plane crashed and burned in a pineapple field. The fire was controlled by a single engine company, and the pilot's charred body was brought back to the firehouse to await the coroner.

In one of the ironies of that day, Lopez' company somehow missed much action as the Waikiki Engine chased down what turned out to be one false alarm after another phoned in by panicky citizens.

Action slowed that afternoon, but firefighters remained jittery as did all of Honolulu as they awaited the return of more Japanese attackers.

"Around sundown we were returned to our Waikiki firehouse," said Lopez. "Two territory guardsmen were assigned to protect our station. One patrolled in front of the station; the other in back. Fire captains were issued .45 caliber revolvers and given orders to shoot saboteurs or anyone else who might interfere with firefighting.

A blackout was ordered that night. Headlights of fire apparatus and all other vehicles were painted black with silver-dollar-sized openings. "When we answered alarms that night, mostly for rubbish, we could barely see the road in front of us. Realizing from the tragedy at Hickam that fire apparatus was a prime target, all rigs were soon painted olive drab to camouflage them," said Lopez.

"Rumors spread that night. We heard that Japanese planes were parachuting invasion troops into the mountains north of Honolulu. Around 9 o'clock we heard the Japanese fleet was outside our harbor.

About this time our firehouse guards ordered a car to stop. The motorist didn't hear them. As the guards opened fire, the motorist turned into our station and told us some damn fools were trying to blow his head off. The guards later mistook each other for invaders and got off several rounds at each other."

Referring again to Honolulu Fire Department alarm records, Lopez offered other indications of that tense night:

9:05 p.m. An alarm which turned out to be only a reflection of light, perhaps from the blazes still out of control at Pearl Harbor.

9:07 p.m. An airplane dropped a flare and touched off a flurry of panicy calls reporting enemy parachutists.

9:08 p.m. A Japanese who was burned out of his home was heating water over an open fire in a vacant lot despite the blackout. He was ordered to douse the fire with buckets of water.

9:14 p.m. Someone using a flashlight was the cause of another report of a fire.

Firehouse tensions eased when no land or air invaders appeared by midnight. "As we watched the smoke and glare from the burning ships at Pearl Harbor and what was left of Hickam, the full impact of the realization that we were at war and that all of our lives would somehow be changed by it, finally hit home," says Lopez. "That was a day I'll never forget."

CHAPTER III

We're At War!

The crew of the US Navy Yard tug *YT-146*, the *Hoga*, was off-duty that tranquil Sunday morning, December 7, 1941, in Pearl Harbor, headquarters of the Pacific Fleet. Quartermaster Bob Brown, 28, second in command of the combination tug and fireboat, was sleeping on a cot in the pilothouse, while nine other crewmen were in the aft deckhouse.

The *Hoga*, a Sioux Indian word meaning "fish," was one of around 100 US Navy vessels, including seven battleships, nine cruisers and more than 30 destroyers — half the United States' Pacific Fleet — nestled inside the naval base. The *Hoga* was berthed with several other tugs at Yard Craft Dock, a few hundred yards across the channel from Battleship Row where 11 vessels, including the battleships USS *Arizona* and USS *Nevada,* were moored at quays just off Ford Island. Many of the vessels along the row were two abreast.

Brown and thousands of other Navy, Army, Marine and civilian personnel at Pearl Harbor and surrounding military airfields, were blithely unaware of the death and destruction that was zeroing in on the island of Oahu. Some 220 miles north of Pearl Harbor steamed a Japanese task force of 33 warships, including six aircraft carriers. The first wave of 185 attackers — fighter planes, high-level bombers, dive bombers and torpedo planes — had lifted off from the carriers at 6 a.m. The second wave, bringing the total assault force to 354 aircraft, followed shortly thereafter.

The *Hoga*, a US Navy combination yard tug and fireboat, rushed to the aid of the *USS Nevada* after the battleship was severely damaged and set afire during the Pearl Harbor attack. Mooring to the port bow of the *Nevada*, the *Hoga's* pilot house monitor gun bored into the flames while the tug's fire pumps fed hoselines put aboard the battleship. The *Hoga* pushed the *Nevada* across the channel leading to open sea and helped to beach her at Waipio Point to avoid the threat that the sinking *Nevada* would block the entrance to Pearl Harbor.

U.S.S. Arizona Memorial, National Park Service

The keenly planned multiple assault included air installations on Ford Island, Hickam and Wheeler Fields, Kaneohe Naval Air Station and the Marine Air Base at Ewa. American planes, most of them standing wingtip to wingtip to better guard against sabotage, would be easy prey. Japan's failsafe tactics called for their destruction or damage before they could get off the ground.

Number one on the Japanese hit list was Battleship Row and the Pacific Fleet Headquarters Base inside Pearl Harbor. Japanese intelligence — faulty, as it turned out — said the Pacific Fleet's three aircraft carriers were also sitting duck targets. If Pearl was struck a devastating blow it would surely destroy the United States' naval and air effectiveness in the Pacific before war was officially declared.

At 7:57 a.m., Brown was jolted awake by the sounds of low-flying, fast-approaching aircraft. He thought this was just another routine war games exercise. Looking from the *Hoga's* pilot house, Brown saw the flaming red Rising Sun emblems under the wingtips of Japanese torpedo bombers passing over the *Hoga.* Battleship Row was dead ahead of them. Swooping low — no more than slightly over 100 feet above water level — Brown stood in disbelief as torpedoes dropped from the bellies of the attackers. Swishing across the channel, the torpedoes bored into Battleship Row as explosion upon explosion erupted into billowing clouds of thick, black smoke and flames. Overhead, bombs rained upon Battleship Row as other planes sighted in on their targets.

The *Arizona*, a short distance up the channel from the *Hoga*, swallowed an unknown number of torpedoes and bombs which mercilessly pelted the battlewagon and exploded. Japan's ace bombardier, Petty Officer

Noburo Kanai, in a high-altitude bomber, released a 1760-pound, 16-inch-diameter armor-piercing bomb which hit near the *Arizona's* No. 2 turret gun, burrowed to the forward powder magazine, and exploded. The tremendous blast lifted the *Arizona* and sent a thick, towering cloud of dark reddish-brown smoke jetting thousands of feet into the sky. The *Arizona* sank in 38 feet of water in less than nine minutes. Only 289 of her officers and crew survived.

Other bombs and torpedoes found their targets along Battleship Row. The explosion-shattered USS *California* began listing and settled to the bottom of Pearl Harbor. The USS *West Virginia* was destroyed by torpedoes and two bombs. With flames swarming over her superstructure, the *West Virginia* followed the *California* and the *Arizona* to the bottom. The torpedoed USS *Utah*, a state-of-the-art, remote-controlled target ship and gunnery training vessel, capsized with the loss of around 63 officers and crew. It would never be raised.

The *Nevada*, tucked at the northern end of Battleship Row, was both lucky and unlucky: Lucky, because massive clouds of black smoke boiling from the *Arizona*, immediately ahead of her bow, obscured the *Nevada* from the attackers; unlucky, because flames floating on top of the water, as burning oil from the hemorrhaging *Arizona's* 9360 tons of oil spilled from her slashed belly, endangered the *Nevada*. Too, the hail of bombs was not selective. The *Nevada* was hit repeatedly, burst into flames in her forepart, and suffered many casualties. The *Nevada* may have been hobbled by the bombs, but she was still able to fight back. Her gunners would be credited with downing two planes.

Chief Boatswain's Mate Joe B. McManus, 31, the *Hoga's* skipper, was aft in his captain's quarters and shaving when the attack started. He ran out on deck and looked across Battleship Row where the USS *Oklahoma*, hit by five torpedoes, was starting to roll over on her starboard side with the loss of 315 officers and crew. Marion Minehart, the *Hoga's* chief machinist's mate, jumped onto Yard Craft Dock to try to see and comprehend what he refused to believe. "Good God!" he said. "We're at war!"

The Yard Craft dockmaster ordered the *Hoga* to get underway, and, according to her log, "assist all ships and pick up survivors." The hours ahead not only challenged the crew of the *Hoga*, but put them in harm's way as undreamed of demands were made. In its favor was the fact that the *Hoga* was just under a year old. Launched December 31, 1940, at Consolidated Shipbuilding Corporation, Morris Heights, New Jersey, McManus and Brown were among those who brought the *Hoga* to Pearl as part of a convoy which encountered turbulently high seas that nearly capsized the boat.

The *Hoga* went in service at Pearl's 14th Naval District Headquarters, May 22, 1941, less than seven months before the surprise attack. Until then, the *Hoga's* mundane duties included assisting ships in and out of Pearl and towing and helping to dock vessels around the 12 miles of the base's facilities.

The welded steel and diesel-powered *Hoga* displaced 350 tons and was registered at 99.7 feet long, with a 25 1/2 foot beam and a 10 1/2 foot draft. The *Hoga*, like many Navy tugs, was outfitted for firefighting, but not for the monstrous fires burning that day. Twin 250 horsepower electrically operated pumps gave the *Hoga* a 4000 gpm capacity at 150 pounds per square inch. Its firefighting equipment included three monitors, one of them atop the pilot house, and four on-deck manifolds on

Smoke barrels from the *U.S.S. Arizona* on Battleship Row. Just ahead of her, left to right, are the sinking *U.S.S. West Virginia* and the slightly damaged *U.S.S. Tennessee.*

U.S.S. Arizona Memorial National Park Service

the starboard quarter of the deckhouse for coupling to firehoses carried by the *Hoga*.

As the *Hoga* backed into the channel that morning, Brown said, "It's a madhouse out there!" His memory is as vivid today — 50 years later — as it was that day the firefighting tugboat went to war. Battleship Row was a maelstrom of explosions, rocketing clouds of impenetrable black smoke and balls of fire erupting from the warships. Tens of thousands of gallons of oil bleeding from the stricken warships spread gigantic smears into the channel where they flashed into flames surrounding ships along Battleship Row.

Carnage was everywhere as explosions flung bodies hundreds of yards onto land, into the water and onto the decks of other ships. As more bombs splattered Battleship Row, other missiles, missing their targets, exploded in the channel and scooped waterspouts spewing hundreds of feet into the air.

Into this unbelievable hell churned the *Hoga*. McManus saw they must immediately help the repair ship, USS *Vestal*, which had been moored next to the *Arizona*. The *Hoga's* crew got a towline on the *Vestal*, damaged by two bombs and her hull shattered by flying debris from the *Arizona*. McManus could not help a moment's reflection upon his recent assignment aboard the *Vestal* and the promotion which resulted in his transfer to the *Hoga*. The repair ship was hauled to the nearby Aiea Bay mudflats where the *Vestal* was beached.

Many bombs hurtling upon Pearl exploded as close as 150 feet from the *Hoga*. It seemed ironic that the warships in Pearl had the firepower to strike back, and many of them were firing upon the attackers. But the *Hoga's* armament shot only water, and one of the crew carried sidearms. They found some comfort in the realization that a tugboat, inconsequential compared to the enemy's primary targets, was unlikely to be singled out for attack. While the *Hoga* was not a priority target for the two waves of attackers with their bombs and strafings, the fact of the matter was that bombs, machine gun fire and falling shell fragments were often indiscriminate and could easily hit the *Hoga,* sink her and slaughter her crew.

The *Hoga* continued her work under the ominous dronings of still more attack planes as the warships mounted counterattacks with machine guns and five-inch shells. The sky over the *Hoga* was pimpled with puffs of black smoke marking exploding antiaircraft shells. The Japanese lost 29 planes during the two-hour attack; a relatively small price to pay for the devastation including eight battleships damaged or sunk; three light cruisers and three destroyers racked by bombs or torpedoes, plus 188 Army, Marine and Navy aircraft destroyed.

At full-ahead 14 knots, the *Hoga* raced to the aid of the *Oglala* and helped the damaged minelayer into 1010 Dock. But it was too late for the *Oglala.* The old, wooden-hulled minelayer capsized. Heading out again toward the burning and smoke-shrouded Battleship Row, where a steady chatter of shells streaked skyward, the *Hoga's* crew spotted two sailors bobbing in the channel. They had either jumped or were blown off ships. One of them suffered a gaping wound in his leg. The other appeared unhurt. The *Hoga* put both survivors ashore at Berth 3 where medical aid awaited them.

Despite its severe damage, fires and casualties, the *Nevada* got underway 40 minutes after the first attack. Under normal conditions it would have taken around two hours to get up steam and be assisted by tugs as it headed to sea. To get there, the crippled *Nevada* would have to run a gauntlet of bombs, torpedoes and strafing as it lumbered, its bow settling lower as it took on water, from the northernmost end of Battleship Row, down the channel and out the narrow opening leading to the ocean. Passing about 50 feet from the blazing *Arizona,* the *Nevada* picked up two survivors, while its gunners shielded their shells with their bodies out of fear that the ammunition would explode from radiant heat.

Japanese attackers immediately saw that one of their primary targets was escaping. Twenty-one planes attacked the *Nevada* with a vengeance that seemed worse than their assaults upon the other warships along Battleship Row. A torpedo exploded amidship on the *Nevada's* port side and blew a hole "large enough to drive two big trucks through," recalled Quartermaster Second Class Roy G. Johnston, 21, who, as assistant navigator, was surrounded by thick steel in the battleship's conning tower. As it made its desperate run down the channel, the *Nevada's* crew fought back with their five-inchers and machine guns while repeated strafings exacted a heavy toll among her crew.

The damage assessment aboard the *Nevada* was ominous. In addition to the torpedo, the battleship had been hit by five bombs. One struck the forecastle, exploded on the main armory deck, backfired and blew up forward of No. 1 turret gun, said Johnson. A direct hit by an incendiary bomb melted and destroyed the entire bridge structure. Fires were raging below deck. The battleship was sinking by the bow. Casualties would include 50 officers and crew killed and at least 109 wounded.

Miraculously, the *Nevada* ran the gauntlet, but was immediately confronted by a new peril. Listing badly and its navigational equipment mostly demolished, the *Nevada* would not be able to make the turn that would take it out of Pearl Harbor and into open sea. If the *Nevada* went down in that narrow channel, it could become not only a navigational hazard blocking the entrance to the Pacific Fleet's headquarters base, but an easier target. The only alternative was to beach her, which was done shortly after 9 a.m., at Landing C, Hospital Point, where many of her casualties could be put ashore.

The *Hoga,* a quarter of a mile north of the *Nevada* and just clearing 1010 Dock, saw the battleship's plight. McManus rang for full speed ahead as the ferocity of the Japanese attacks continued. Glancing up, McManus saw a triangular pattern of bombs hurtling down dead ahead of the *Hoga*. He put the boat full astern as the missiles plopped into the channel, but failed to explode. Still the Japanese refused to be denied the *Nevada.*

As the *Hoga* closed upon the battleship, a bomb, believed to have been intended for the *Nevada,* struck the forward deck of the destroyer USS *Shaw* in Floating Drydock No. 2. The *Shaw's* ammunition stores exploded, wreathing the destroyer in flames and smoke crowned by flaming trails of skyrocketing shells.

By 9:25, according to the *Hoga's* journal, she had put her port bow alongside the port bow of the grounded *Nevada,* whose on-board firefighting system had been blasted away. The *Hoga* switched to firefighting mode. A stream jetted from the monitor on the pilot house into the flames while *Hoga* crewmen coupled four hoses to its discharge outlets. The *Nevada's* crew pulled the hoselines on board and attacked the fires. Intense smoke and flames below deck involved storerooms and bedding

View of the *U.S.S. West Virginia* and *Tennessee* on Battleship Row after the attack.

U.S.S. Arizona Memorial
National Park Service

which coughed flame and smoke from bomb holes and open hatches in the forward section. The *Nevada's* hull rapidly heated until layers of gray paint on the hull began to burn.

Still the fires worsened. The *Nevada's* crew called for more hose. The *Hoga* had none to give. "Better start thinking about forming a bucket brigade," McManus called to them. With the *Hoga* secured to the *Nevada,* McManus watched with growing worry as the anchor continued to settle until it was directly opposite him. The *Hoga's* pilot house and top-mounted monitor gun soon stood well above the *Nevada's* bow. There was no question that the battleship was going down at the strategically worst possible place: the entrance to Pearl Harbor.

A clever plan was devised, recalled Johnson. The *Hoga* would help push the *Nevada* backward across the channel where it would be beached, stern first. That would not only clear the sea lane, but point the *Nevada's* guns out to sea. As a beached fortress, the *Nevada* would serve as a deadly warning to Japanese submarines and other warships reported to be preparing to invade Pearl Harbor. Unknown to the enemy, of course, was the fact that the guns of the *Nevada* were useless because of the heavy damage she had suffered.

The tiny *Hoga,* dwarfed by the *Nevada* which was nearly six times longer than the tug, began pushing the battlewagon across the channel while its pilot house monitor gun continued to bore into the flames. The cross-channel maneuver was a navigational feat that would long be remembered because it was photographed for posterity by a Navy photographer. With the *Nevada's* engines reportedly at one-third speed astern and the

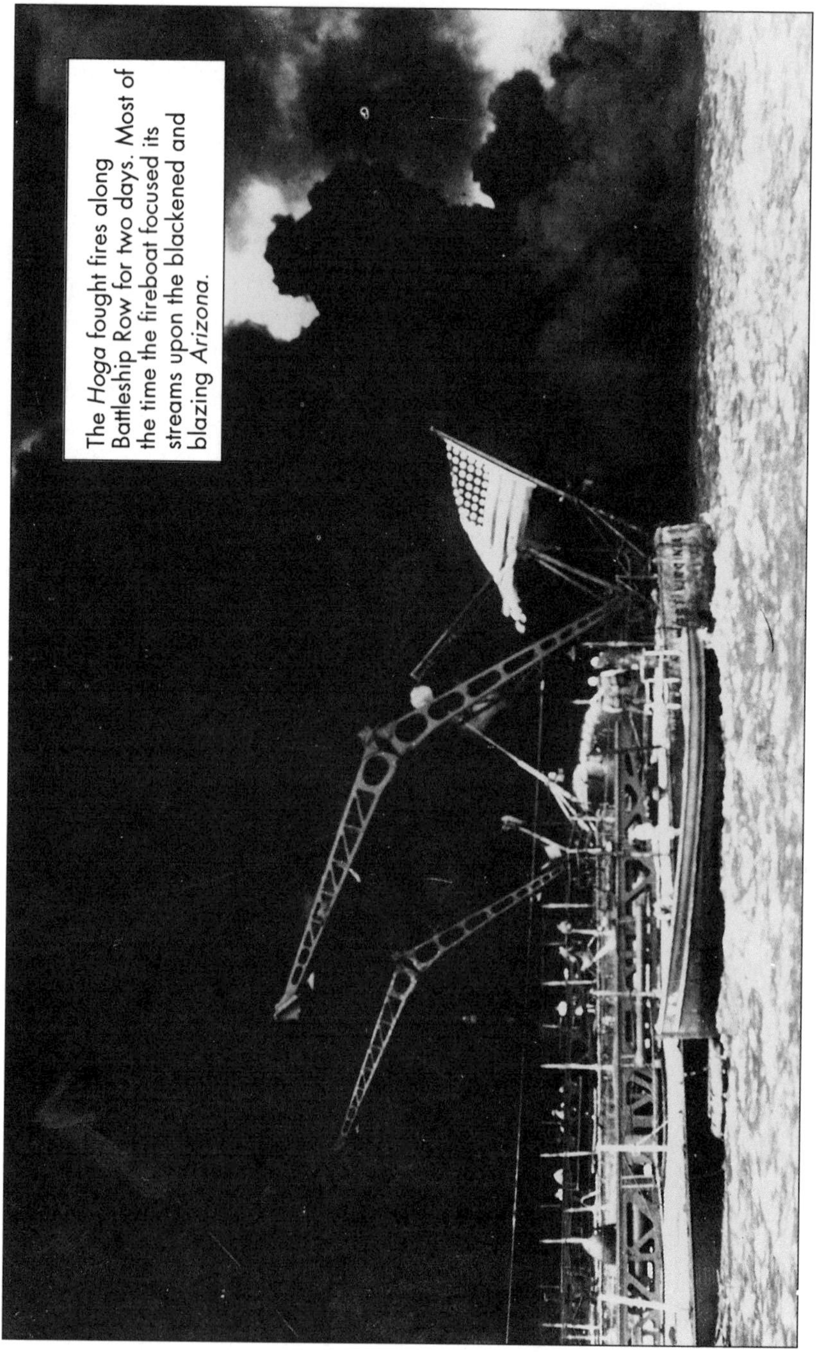

The *Hoga* fought fires along Battleship Row for two days. Most of the time the fireboat focused its streams upon the blackened and blazing *Arizona*.

Hoga pushing full ahead, the channel crossing seemed to take forever. Just before the *Nevada* was grounded near a Waipio Point sugar cane field, her engines were stopped. The *Hoga* gave a mighty push, and the battleship's stern went solidly aground as the anchor was dropped. The *Nevada* was no longer a navigational threat to other vessels.

Firefighting continued. At 12:40, according to her log, the seaplane tender *Avocet* arrived to lend more firefighting assistance. Fires were not controlled until 2:30 p.m. With the *Avocet* continuing to help, the greater firefighting capabilities of the *Hoga* were needed along Battleship Row. The *Hoga's* streams began boring into flames raging inside the blackened *Arizona*. The *Hoga* remained at the *Arizona's* side for two days. The fire tugboat's crew saw, but could not help, much less recover, many of the battleship's dead and drowned officers and crew.

At 1 p.m., December 9, two days after the sneak attack, the *Hoga* was relieved of its firefighting duties and returned to its berth at Yard Craft Dock to await further orders. The following days, according to the *Hoga's* log, included searching for bodies, removal of damaged ships and debris from the channel, and patrol searches for reported enemy submarines.

Almost constantly in the line of fire during the two-hour attack, none of the *Hoga's* crew was injured, nor was the boat damaged. Somehow the story started that a large dent on the front port quarter of the *Hoga* is a memento of pushing the *Nevada* onto a spot which today is known as Nevada Point. McManus and Brown continued to insist that the crease in the hull never occurred during their time on the *Hoga*.

The *Nevada* was raised and refloated, February 12, 1942, several months after the surprise attack. Following a $23 million reconstruction and modernization at the Puget Sound Navy Yard, Bremerton, Washington, she rejoined the fleet. The *Nevada* saw much World War II action, including the invasions of France, Iwo Jima and Okinawa. Later, she had the ironic experience of participating in the occupation of Japan when the war ended. The *Nevada* was used as an experimental vessel to test radiation during the July, 1946, atomic bomb tests at Bikini. Decommissioned that year, the *Nevada* was used as a target by other warships testing new armament. At 2:30 p.m., July 31, 1948, gunfire and aerial torpedoes sent her to the bottom of the ocean off Hawaii.

The *Arizona,* still lying where she sank, was dedicated on Memorial Day, 1962, as the USS *Arizona* Memorial. The more than 1100 Navy and Marine Corps personnel who went down with her are forever entombed in her hull. A total of 2403 military personnel and civilians were killed in the surprise attack. Bombardier Kanai, whose mammoth bomb sank the *Arizona,* was shot down and killed during the Wake Island battle shortly after Pearl Harbor.

Not until years later did McManus and Brown learn of the *Hoga* citation by Admiral Chester W. Nimitz. As commander-in-chief of the Pacific Fleet, Nimitz officially recognized the *Hoga's* crew "for distinguished service in the line of your profession . . . efficient action and disregard for your own personal safety . . . . when another ship *(Nevada)* was disabled and appeared to be out of control, with serious fires in the forepart of that ship, you moored your tug to her bow and assisted materially in the beaching operations in an outstanding manner. Furthermore, each member of the crew of the *Hoga* functioned in a most efficient manner and exhibited commendable disregard of personal danger throughout the operations."

Seven years after the Pearl Harbor attack, the *Hoga* began a new mission when it was loaned by the Navy to the Port of Oakland, California, to serve as that city's first fireboat. Following the official transfer of papers, May 28, 1948, more than $73,000 was spent by Oakland to increase the boat's capacity from 4000 to 10,000 gallons-per-minute at 150 pounds per-square-inch to meet the needs of the rapidly growing port. Upgrading was done by the Pacific Coast Engineering Company at Pacific Drydock and Repair in Oakland.

Retained were the *Hoga's* twin McIntosh & Seymour (American Locomotive) of Auburn, New York, 650 horsepower at 740 rpm main diesel engines which operated two 419 kilowatt Westinghouse electric generators for propulsion or pumping requirements. The boat's pumping capacity was increased by the addition of three United States Iron Works pumps, each single-stage contrifugals. They were powered by three 250 horsepower supercharged Buda diesels. One of these was later replaced by a D333 Caterpillar.

Because there was insufficient engine room space for these additional pumps, they were installed on the main deck abaft the deckhouse. This above and below deck pump location is believed to be unique among fireboats. The weight of these aft-mounted pumps largely resulted in a new draft for the boat: nine feet forward and twelve feet aft. As rebuilt, the boat displaces 343 net tons.

Beginning of salvage operations on the *Nevada*, which was raised and refloated several months after the attack. She participated in the Iwo Jima and Okinawa battles.

U.S.S. Arizona Memorial National Park Service

The *Hoga* was loaned by the US Navy to the Port of Oakland after World War Two. The combination yard tug and fireboat was converted into a single-purpose fireboat. While the boat retained most of its original configuration, its pumping capacity was raised from 3000 to 10,000 gallons-per-minute at 150 pounds per-square-inch. Note the unique outboard water supply main that encircles the deckhouse. The original pilot house monitor, used to fight fires resulting from the Pearl Harbor attack remained in use, but the four hose discharge outlets on the deckhouse did not.

David C. Henley

Also part of the upgrading was the installation of additional turret guns. Three were located port, starboard and center at the end of the upper deck platform. The fourth was mounted at the stern. The original tower-mounted gun is elevated about 14 feet above the upper deck. Another gun is mounted on the pilot house lip. In 1989, the *City of Oakland* had six monitor guns.

The guns and twelve gates hose discharge outlets, evenly distributed on the port and starboard sides of the deckhouse, are supplied by a common loop, or main. This loop encircles the upper area of the deckhouse. A series of fog nozzles, mounted immediately above the loop and

When Pearl Harbor's *Hoga* was loaned to Oakland in 1948, the city increased the boat's pumping capacity to 10,000 gallons-per-minute at 150 pounds per-square-inch. On December 2, 1948, the day after the boat went in service, it was called to flood the hold of the burning *Hawaiian Rancher*.

David C. Henley

supplied by it, provides protective sprays against radiant heat. By 1989, four foam-discharging hose outlets were fitted under the pilot house lip. The boat carried 200 gallons of high-expansion foam.

Renamed the *Port of Oakland* (changed to *City of Oakland* after 1955) the fireboat went into service December 1, 1948. And none too soon. The following day the fireboat met the *Hawaiian Rancher,* a Matson freighter inbound from Stockton, California. The captain had reported a fire in No. 2 hold as he came into San Francisco Bay and approached the San Francisco-Oakland (Transbay) Bridge.

The *Port of Oakland* escorted the ship into the Encinal Terminal, Alameda. Intense heat and smoke forced flooding of the hold with hoselines supplied by the *Port of Oakland.* The fireboat narrowly escaped severe damage or destruction when the *Hawaiian Rancher* suddenly listed and seemed about to capsize.

A year later the *Port of Oakland* pumped for more than nine hours as it delivered more than six million gallons of water while battling the Pier 4, Oakland Army Base, Outer Harbor, fire, March 15, 1949. The fire was, at that time, the largest and most destructive in the history of the Oakland Fire Department.

Considering her historic fame, the *City of Oakland* was the obvious choice to take President Jimmy Carter on a July 3, 1980, tour of the port. The crew presented him with a copy of the famous photo of the boat fighting fires aboard the *Nevada.*

These are only 54 of the principal German cities which were set aflame by Allied bombers.

CHAPTER IV

An Eye For an Eye

With the United States' declaration of war against Germany, Italy and Japan, the Axis powers soon learned that American and British warplanes would, with their own terrible swift swords, retaliate in ways never before seen in the history of mankind.

Payback time was not as immediate as many hoped, but long-range planning more than justified these delays. High on the hit list were Berlin, Hamburg, Cologne, Dresden, Stuttgart and other strategic targets. The final and worst attacks by the German Luftwaffe on London, December 29-30, 1940, and May 10, 1941, provided the basic ingredients for Allied planning.

Firemen and strategists knew, if the Germans did not, that the December and May precision bombings started fires and caused destruction in relatively small sections of London. Firefighters were overwhelmed. Without adequate water, fires rapidly escalated into uncontrollable cauldrons. Noted, too, were the large numbers of firemen killed, a factor which would be taken into consideration when German cities were attacked.

The Luftwaffe had, moreover, slugged London with around 500 planes during each raid. What would result, strategists wondered, if the Allies doubled the number of attack planes to 1000? More planes and larger planes would carry heavier bomb loads. Instead of shotgunning their attacks over large areas as the Germans did until those December and May attacks, Allied planes would saturate specific areas. What crop of fires and destruction would they harvest?

How much more havoc would be caused from virtually around-the- clock bombings? Planners recalled that the Royal Air Force had barred the Luftwaffe from further daytime raiding about a week after the start of the London Blitz. From that time forward, German attackers would be forced to fly at night.

London firefighters had at least a chance to control fires and regroup during daylight hours. Theorists were fairly certain that day and night bombings would wreak heavy damage while overwhelming German firefighting forces. The chances of success were better due to the Luftwaffe being stretched thinly over the Russian and other fronts.

Also at issue was the question whether Germany's resolve would equal that of their British neighbors across the English Channel. How would civilian populations react to morale-shattering attacks on cathedrals and other historic landmarks?

There was one question that would not be answered until after the war: Why did incendiaries fail to set Berlin afire? That puzzle would be solved after the war by the legendary Chicago firefighter, Chief Fire Marshal Anthony J. Mullaney.

The old city of Cologne on the Rhine River was among the first hit. Cologne was the largest shipping center in the northwestern part of the country and the nation's cradle of German culture. In May, 1942, some 1000 Allied planes pounded Cologne with 1500 tons of bombs in less than 90 minutes.

U.S. Eighth Air Force B-24 Liberators destroyed this oil refinery on the Rhine River near Cologne.

German soldiers stare forlornly at blazing ruins of the Fieseler aircraft plant at Bettenhausen following Allied incendiary bomb attacks.

A gothic cathedral, whose twin spires pierced the skyline for more than 500 feet, was hit. Similarly, other churches dating back hundreds of years were shown no mercy. The renaissance-era City Hall and other government buildings were hit. Smoke clouds billowed three miles into the air from fiercely burning fires inside the bombed out shells of Cologne's landmarks.

Cologne's devastation caused longstanding bitterness among Germans who considered the raid inexcusably vindictive. The city was, true enough, an irreplaceable treasure chest of history. On the other hand, bombs are not choosey, although nobody doubts that the Allied High Command targeted cathedrals. Cologne's commercial importance, moreover, made it a legitimate target.

Hamburg on the Elbe River near the North Sea was Germany's largest seaport. Its strategic value to the war effort was profound. Docks, wharves, warehouses, shipyards, heavy machinery and chemical manufacturing plants blended from the crowded waterfront into even more tightly-compact areas of cathedrals, tenements and apartment buildings. Here had been the homes of Brahms and Mendelssohn.

Authorities perhaps had cause to be confident because of the relatively low incidence of major fires. Most buildings were built of sturdy brick and masonry walls with tile or slate roofs and other built-in fire resistant features. Rarely were structures of wood frame construction.

There were, however, fusible links in these allegedly fireproof buildings. Roofs, mostly peaked at right angles to streets, were supported by thick wooden sheeting and criss-crossed underneath by networks of timber supports. From the exterior, Hamburg buildings may have been relatively safe. Inside they were not. The city had but two known buildings outfitted with automatic sprinkler systems. Neither would suffer serious fire damage.

Hamburg was ripe for burning on the clear night of July 24, 1943. Intelligence sources reported unusually high temperatures and low humidity readings that preheated the city. Shortly before midnight, American and British Liberator and Flying Fortress bombers began to sledgehammer Hamburg and continued without letup for five days.

Thousands of tons of high-explosive and incendiary bombs showered Hamburg. Extremely awesome in their deadliness were M-69 incendiaries. Each weighed slightly over six pounds and plummeted 225-feet-per-second. Like knives cutting through warm butter, they easily pierced slate and tile roofs. Exploding in attics, each bomb's canister detonated over six pounds of jellied gasoline while liberating 13,000 units of heat. Dropped in clusters of three dozen, the results were horrendous.

Hamburg became an almost instant conflagration.

Approximately 54 miles of buildings erupted in flames inside the targeted zone. With no exaggeration: Hamburg became an almost instant conflagration. Over 16,000 blazing buildings along the waterfront and in residential districts fed the inferno. More than 60,000 people perished in just one of the first major raids. A German professor and consulting pathologist who survived Hamburg wrote:

The central district of Hamburg lay in ruins following late July, 1945, attacks by Allied bombers.

"The crowded conditions in the city with its few parks or large squares, the height of apartment houses and the age of the dwellings are all contributory factors to the magnitude of the catastrophe. Incendiary bombs started fires which spread particularly in thickly-inhabited parts of the town in a very short period of time.

"In several minutes whole blocks were on fire and streets were made impassable by flames. The heat increased rapidly and produced a wind which soon was of the power and strength of a typhoon. This typhoon first moved in the direction of the fires, later spreading in all directions. In the public squares and parks it broke trees... Burning branches shot through the air. Trees of all sizes were uprooted.

"The firestorm broke down doors of houses and later the flames crept into the doorways and corridors. The firestorm looked like a blizzard of red snowflakes. More scientifically, a firestorm is a mass of fresh air which breaks into burning areas to replace the superheated rising air. The heat turned whole city blocks into flaming hell. In the course of hours the air in shelters became increasingly worse. "Matches or candles did not burn (for lack of oxygen)." Altogether, 100,000 people would die during the raids; 70-80 percent of them by carbon monoxide poisoning which wormed its way into the very shelters that were supposed to save them.

Altogether, 100,000 people would die during the raids.

Lacking any effective pre-war firefighting plan, Hamburg's efforts to save the city were doomed. Unlike London, where at least 1000 fire engines quickly answered calls to Blitz fires, Hamburg was only able to muster 86 units during the early hours of the first raids. German firefighting resources were a hodge-podge of paid and volunteer forces backed by the military with self-contained mobile fire units and airborne firemen.

On paper, the system was severely flawed. In practice, German military requirements drained local firefighting forces. Women and Hitler Youth Groups were drafted as fire personnel. Not only was training or hands-on firefighting experience inadequate or nonexistent, but incoming troops were not familiar with Hamburg's streets, special hazards, water supply locations and all the other ingredients necessary for efficient firefighting.

Fires over the years had lulled Hamburg fire officials into a continuation of traditional tactics. They quickly realized their blunders and never recovered from them. In tried-and-true ways they answered alarms from their fire stations. Fighter planes strafed firemen and their equipment.

A new approach also proved futile: Firemen rushed from their stations to shelters until air raid sirens sounded the all-clear. Only then did they waste additional time getting to their firehouses and heading out to fight fires. Escort fighter pilots snuffed that ploy by returning and strafing firemen and their apparatus. Firemen became as ducks-in-a-barrel and were slaughtered by machine-gun fire as falling walls buried them and their apparatus.

A third attempt to outwit attackers similarly failed. Firemen moved their apparatus to outlying forests and took cover until raiders went home. By then, whole areas of Hamburg were burning and blending into unstoppable walls of fire.

Hamburg's fire apparatus was burned and buried by debris in this street when firemen were forced by heat to abandon their rigs as a firestorm turned thoroughfares into chimney flues filled with flames and flying brands.

As military and civilian fire companies hurried to Hamburg they were greeted at the outskirts by motorcycle and bicycle messengers who gave them maps showing where they were to report. By the time they reached their designated positions, flames had long since overrun what they were supposed to protect. Without radios, fire chiefs were unable to coordinate and mount any sort of attack, much less change tactics as conditions demanded. It was not a matter of surrounding and drowning fires with water. Hamburg's apparatus and water supplies were never intended to fight huge fires.

The fire department's first line of defense consisted of pitifully inadequate 660-gallons-per-minute pumpers. These might put out a small fire, but could not stop flames raging along streets and swirling among ships, warehouses and factories on the waterfront.

Water supplies were similarly feeble and had not been well-maintained. Corrosion and sediment in the incredibly small mains forced pump operators to draw from three hydrants at a time, rather than the customary one. To avoid broken hose, pump pressures were limited to only 75 pounds-per-square-inch.

Firehose was less than 3-inches in diameter, far smaller than flows required in nearly all American communities,especially high-hazard areas. Hand-held nozzle tips were only slightly larger in size than that of an ordinary garden hose. Hamburg's fireboat nozzles were not much better. In the maelstrom of smoke and fire that was the waterfront, the fireboat was a mere squirt gun.

For lack of trained and inexperienced personnel, woefully inadequate apparatus, hose and water sup-

The small-sized nozzles of Hamburg's fireboat were mere water pistols when it came to fighting fires along the waterfront.

plies, Hamburg firemen were, as Churchill called British firefighters: "Heroes With Grimy Faces." But few of them survived to bask in any glory, much less medals or monuments. To their everlasting credit, Hamburg firemen put out their best efforts with what little they had to fight fires and rescue whoever survived. Probably no firemen nor any fire department in the world could have done more.

Hitler seemed more determined to protect his battlefronts than his homefronts. Did he not realize that his armies, navy and air force depended upon Germany's industrial capacity? And was he not more obsessed with conquest than caring one whit about the German people or their centuries-old culture as evidenced by cathedrals, museums, art, literary and music traditions?

They called Hamburg "The Dead City' following raids which obliterated residential, commercial and waterfront districts.

The country that called Bach, Goethe, Beethoven, Brahms and Haydn native sons was on a sure-fire path to destruction. If any more proof of Hitler's paranoia is needed, the millions of innocent men, women and children who perished during the Holocaust need only be cited.

Starting February 13, 1945, and continuing for two days, the nightmares of Hamburg gave way to catastrophic high-explosive and incendiary attacks upon Dresden, an industrial and cultural center in eastern Germany. Raiders wiped out heavy industrial complexes, machine tool plants and production of optical instruments.

Dresden, established before Columbus' arrival in America, suffered far more lethal blows than Hamburg and in a shorter period of time. City Hall, the Zwingen Palace and Museum, together with the city's major

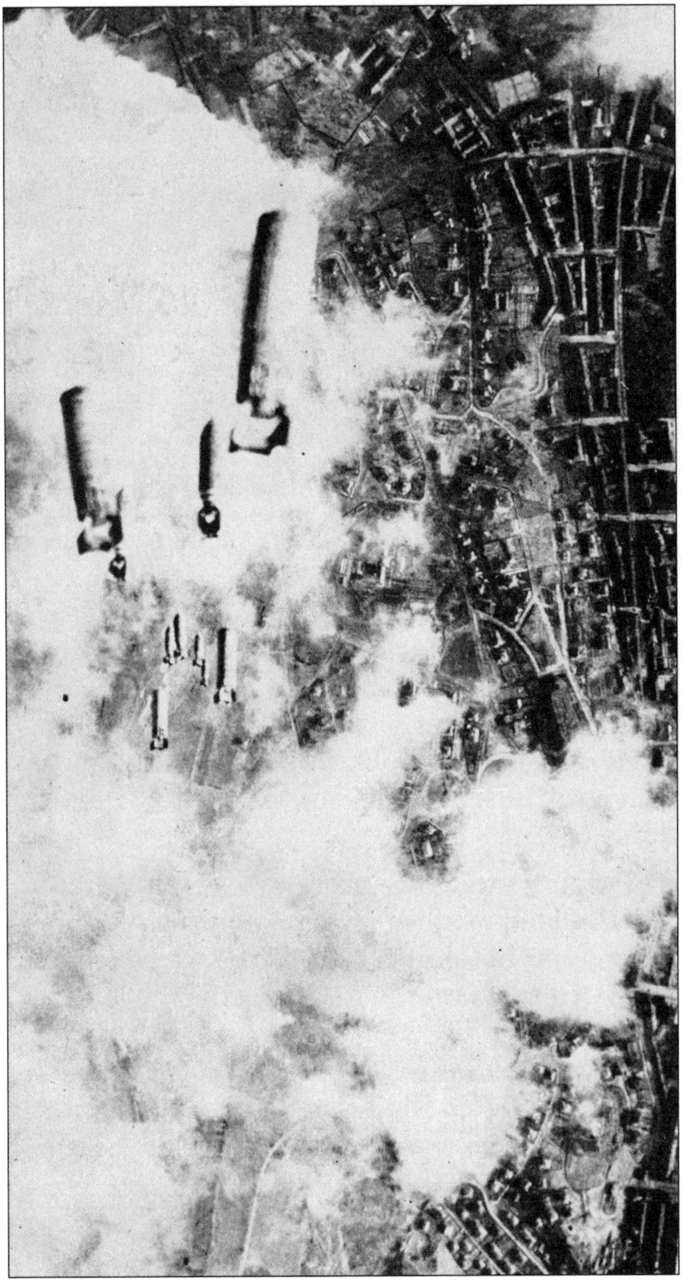

St. Valentine's Day, 1945, heavy incendiary attack on Dresden by U.S. Eighth Air Force bombers sets additional fires after British Royal Air Force planes pounded the city the previous night.

Berlin before (left) and after (right) aerial photos of the city's Tiergarten district. Stereoscopic lenses provided three-dimensional comparisons. The question of why Berlin did not burn is answered in the aftermath photo (right) showing how fire division walls, rarely seen in other cities attacked, prevented firespreads.

cathedral took heavy blows as thousands of clusters of incendiaries set the city aflame. Casualty figures vary, but between 135,000 and 300,000 residents were killed, bringing to more than 500,000 people dead throughout Germany until Hitler surrendered less than three months later, May 7, 1945.

The significance of the air raids was summarized by Chief Mullaney: "Allied air forces produced the intense firestorms in German cities which were perhaps the most important factor in producing the civilian and military collapse of the German Nation." In assessing areal photos taken after the Berlin raids Mullaney discovered why Berlin did not burn. He detected the parapeted fire-stopping walls that honeycombed Berlin and saved it.

In the United States, meanwhile, war placed heavy burdens upon the nation's seaports through which war supplies were funneled to the European and Pacific theaters. Ship traffic surged with troop embarkations and war materials handling. More cargoes and more ships meant potentially more fires, especially with the large volume of petroleum products, ammunition and other explosives. Sabotage was a constant fear as was enemy air attack and even invasion. If Pearl Harbor could be devastated by surprise, why not Los Angeles, New York, and other cities.

Wartime experience showed the fire service had more to fear from carelessness in ports than a saboteur's bomb. Peacetime fire safety regulations were often ignored. The 1942 loss of the French ocean liner, *Normandie,* was a prime example. This world's largest luxury liner was being converted into a troopship at New York's Pier 88 on the North River at the foot of West 45th Street. A welder's torch is believed to have ignited kapok-filled life preservers. Instead of immediately turning in an alarm, workmen fought the rapidly worsening fire for at least a quarter of an hour.

At 2:49 p.m., February 9, 1942, the first of five alarms was sounded from Box 852 for the burning *Normandie.* Fireboats and hundreds of firefighters from land-based companies poured millions of gallons of water in the liner as fire gnawed deeper in the *Normandie.* The fight to save the $50 million vessel continued through the night. The following day the *Normandie* slowly rolled over on her port side while the *Fire Fighter,* refusing to quit, continued to pour water into a ship that was lost.

Although the $50 million *Normandie*, the world's largest luxury liner at the time, lays capsized and ruined, New York's fireboat *Fire Fighter* refuses to quit the battle to save her. The *Normandie* was being converted into a troop carrier when a welder's torch is believed to have ignited a fire on February 9, 1942. Before calling the fire department, workers and supervisors fought the worsening fire with a scenario which the National Fire Protection Association described as "a Hollywood slapstick comedy." When the alarm was finally turned in, hundreds of firefighters from land and fireboat companies were unable to prevent the loss of the *Normandie* to the war effort. The liner capsized the following day.

From the Collection of Paul Ditzel

Germany celebrated what their propagandists claimed was the successful sabotage of the *Normandie*. The National Fire Protection Association's report proved otherwise. Sharply critical of the negligence and ineptitude of the ship's workers and their supervisors, the NFPA report said their efforts had "the elements of a Hollywood slapstick comedy" before they thought to call the fire department.

War concerns were put aside in Boston on Saturday, November 28, 1942, as hundreds of servicemen about to be shipped overseas turned out for the Boston College-Holy Cross football game at Fenway Park. The winner would play in the Sugar Bowl. The Boston Globe's sports columnist wrote, "This is anything can happen day." Holy Cross' underdog team won, 55-12.

Among the fans was Buck Jones, the famous Hollywood film cowboy, who later joined 1000 others that night in the Cocoanut Grove, a Prohibition-era speakeasy which had been converted from a garage into Boston's best-known nightclub at 17 Piedmont Street.

At 10:20 p.m., Boston firemen answered an alarm for a burning automobile near the Cocoanut Grove. After dousing the fire they were returning to their stations when firemen saw puffs of smoke rolling from the nightclub. At that moment, a passerby pulled Fire Alarm Box 1521 close to the Cocoanut Grove. Arriving on Piedmont Street, firemen were met by hundreds fleeing from the nightclub where an artificial palm tree was ignited in the basement Melody Lounge, perhaps accidentally by a busboy.

Four more alarms were quickly sent in as firemen got 18 streams operating. Locked exits, a blocked revolving door, highly-flammable decorations and air conditioning fans created a flash-fire that, inside of five minutes, sent thick toxic smoke throughout the building. Within half-an-hour 498 people, including Buck Jones, were killed and nearly 200 burned or poisoned by the carbon monoxide-laden smoke during the worst nightclub fire in American history.

Nearly 500 people, including many servicemen about to be shipped overseas, perished during a flash fire, November 28, 1942, in Boston's Cocoanut Grove nightclub.

A year later, the Port of New York area was again threatened by catastrophe when fire broke out after a fuel pipe burst in the boiler room of the Panamanian freighter, *El Estero,* berthed at the U.S. Army Caven Point Terminal in Bayonne, south of Jersey City. The *El Estero* had 1400 tons of high explosives on board and was loading high-octane aviation gasoline when the fire was discovered.

Perhaps remembering the *Normandie,* an alarm was immediately turned in at 5:30 that Saturday afternoon, April 24, 1943. The deep-seated fire quickly spread toward the ammunition cargo holds and gasoline compartments of the *El Estero,* berthed in an area studded with clusters of storage tanks containing millions of gallons of petroleum products.

Despite efforts of Jersey City firefighters, the situation was worsening an hour after the first alarm, even with New York's two largest fireboats the *Fire Fighter* and the *John J. Harvey,* tied up alongside the freighter's port and starboard sides, continued to cascade tens of thousands of gallons into her. Under the weight of the water sloshing inside her, the *El Estero* suddenly listed to starboard. Cargo and parts of the ship crashed onto the *Fire Fighter.* Its crew, expecting the *El Estero* was about to capsize, cast off. When it did not, the *Fire Fighter* quickly returned and opened up its deck pipes again. If fireboat streams could not penetrate the heart of the fire, perhaps they could reach the cargo holds and the ship would go down before she exploded.

Around 9 p.m., the *El Estero* was about one-half mile due west of Robbins Reef Lighthouse when the freighter sank. The disaster alert was lifted. New York fireboat firefighters and Jersey City firemen were awarded citations for their heroism in averting an almost certain catastrophe.

On July 6, 1944, exactly one month after the D-Day invasion of Europe, firemen in Hartford, Connecticut, answered an alarm in the city's North End. A small fire had ignited a support rope inside the Big Top of the Ringling Brothers and Barnum & Bailey Circus shortly after the afternoon matinee started.

Karl Wallenda of The Great Wallendas high-wire acrobatic family was about to start his act when he spotted the fire. He called the world-famous clown, Emmett Kelly, who ran for a bucket of water as more than 6000 people, mostly children, watched and wondered whether this was a circus stunt.

What appeared to be a trifling fire suddenly zipped up the rope and flashed into a fully blossoming blaze

One month after the D-Day invasion of Europe, the Big Tent of the Ringling Brothers and Barnum & Bailey Circus turned into a deadly canopy for 168 people, many them children. Hartford, Connecticut firemen did their best, but the stampede for exits and the rapidly spreading fire thwarted them.

that mushroomed across the Big Top. The tent was coated with highly-inflammable paraffin wax and other oils to weather rainstorms. It not only failed to repel the spreading flames but fed them as they streamed across the canvas umbrella.

Panicy children screamed. Parents snatched them and stampeded for exits. Grabbing his cornet, Band Leader Merle Evans began to play "The Stars and Stripes Forever", a pre-arranged signal to circus artists and roustabouts that they were needed for an emergency under the Big Top.

Disaster was occurring even before firemen pulled out of their stations and headed for the thick clouds of smoke chuffing into the hot and humid afternoon sky.

They discovered pandemonium upon their arrival. Quickly they set up hoselines and began to pull out the alive and the dead. Inside of 30 minutes, 168 people, 67 of them children, lay fatally burned or trampled.

> **Inside of 30 minutes, 168 people, 67 of them children, lay fatally burned or trampled.**

"The scene of America's worst catastrophe was a day in which clowns cried and coffins became scarce," wrote Louis G. Black, an Associated Press reporter who covered the calamity.

The Ringling Brothers fire took on a seemingly endless mystique. Authorities were not able to identify a small girl although she was untouched by the flames and seemed peacefully asleep. For years she was known as "Little Miss 1565", after her morgue tag number prior to her burial. Her picture was published in newspapers around the world. The mystery of her identity deepened over the years.

Nearly 50 years later, Lieutenant Rick Davey of the Hartford Fire Marshal's office, chased down clues, tips and false alarms. On March 8, 1991, Lieutenant Davey identified "Little Miss 1565". The body of Eleanor Emily Cook, 9-years-old at the time of the fire, was exhumed and reburied in her family's plot in Southhampton, Massachusetts. Her mother, still alive, had been badly burned in the fire and her older brother had perished. For years, Emily was thought to be one of the six victims who were so badly charred that their identities would forever remain unknown.

The nation was still grieving over victims of the Ringling Brother circus fire when, 11 days later, the United States' worst wartime disaster occurred near the isolated town of Port Chicago, 40 miles north of San Francisco. The Port Chicago U.S. Naval Ammunition Depot was the Pacific Coast's largest bombs and bullets transshipment center.

Two Liberty Ships were berthed alongside the depot's 1200-foot-long pier that night of July 17, 1944. The newly-launched 10,000 ton *Quinault Victory* was taking on munitions destined for the Pacific Theater of war. Hundreds of black sailors labored as stevedore gangs while their white officers watched and made bets with each other on which gang would load the fastest.

A short distance away the similarly-sized *E.A. Bryan* was preparing to sail; its holds filled with more than 4600 tons of bombs. The only fire protection was a U. S. Coast Guard barge moored nearby. It was not only poorly placed if trouble happened, but its crew were firefighters in name only.

At 10:19 p.m., two explosions obliterated the two ammo ships, the Coast Guard firefighting barge and everybody and everything within a radius of 1000 feet. Chunks as large as firehouses hurtled skyward. A U.S. Navy patrol pilot maintaining an 8000-foot altitude broke radio silence and said he was climbing to 12,000 feet to avoid exploding bombs and pieces of the ships.

Miles away, witnesses reported a lightning-like flash followed by glowing red streaks piercing the sky. Shock waves rattled windows in San Francisco's St.

Francis Hotel. The University of California at Berkeley seismograph needles jiggled for three minutes while registering a 3.4 magnitude earthquake.

Port Chicago's power and water supplies were knocked out by the blasts and pieces of the ship thunking onto the town. There were few outbreaks of fire and these were mostly trivial. Fire departments from miles around sent water tankers to supply drinking water. From the San Francisco Fire Department came Searchlight Engine No. 1 to illuminate ground zero of the disaster.

The five-seconds apart explosions killed 320 servicemen, 202 of them black. Cause of the disaster was never determined. One clue was provided by accounts of the speeded up loading ordered by officers. More than

The U.S. Naval Ammunition Depot near Port Chicago, California, was mostly leveled by explosions aboard two Liberty ships. Fire departments for miles around rushed to help.

200 black sailors refused to return to work later unless safeguards were implemented and they received the same furloughs that the white officers did immediately following the disaster. Many were found guilty of mutiny, court-martialed and imprisoned on Terminal Island, San Pedro, California. Ultimately most were freed or pardoned, but congressional attempts to completely clear their names failed as the 21st Century neared.

With Port Chicago quickly back on line, the flow of munitions resumed. Soon would arrive an ugly big blob of a bomb secretly destined for some place called Hiroshima.

The arrow points to the origin of the liquefied natural gas (LNG) explosion at the East Ohio Gas Company plant in Cleveland, Ohio. Twenty-seven engine companies, seven hook and ladders, together with other firemen fought the five-alarm blaze resulting from the 2:43 p.m. blast, October 20, 1944. Destroyed or damaged were 79 homes, two factories, eight mercantile properties and 200 automobiles. The disaster killed 130 people and injured 225, including 23 firemen. The cause was attributed to a metal tank that cracked while holding LNG stored at 257-degrees below zero Fahrenheit.

Bravery was similarly commonplace among fireboat firefighters in Los Angeles during World War II where gross negligence touched off a disastrous explosion and fire. The tankship *Fredericksburg* was at the Union Oil Terminal's Berth 151 that Saturday afternoon, October 21, 1944, and taking on a cargo of toluene, an anti-knock additive for military aircraft fuel. Ship security, if indeed there was any, failed to notice that toluene was leaking from the tanker.

The leak went undetected for at least six hours as a volatile smear fanned out from the *Fredericksburg.* The tide carried this floating bomb across the Turning Basin to Berth 223, Terminal Island. A blanket of toluene surrounded two Navy landing craft moored side by side, while welders prepared them for war. Still seeking ignition, toluene vapors accumulated under a 200-foot-long wooden wharf where Navy and civilian workers had parked their cars and trucks.

Ignition, probably caused by a welder's torch, was followed by sheets of flame and a towering cloud of black smoke. The 1:58 p.m., explosion was seen and heard by firefighters at *Fireboat No. 2,* berthed a few hundred yards south. The fireboat and its auxiliary, *Fireboat No. 3,* arrived in minutes.

As Pilot Brainard "Choppy" Gray brought *Fireboat No. 2* close, Captain Maurice P. Allen ordered the boat's most powerful gun, Big Bertha, to attack. Opening the discharge valve, a fireman directed a 10,200 gpm wallop. With one sweep, Big Bertha knocked down flames aboard both landing craft. A second sweep successfully battered flames on the dock, its crane and 25 vehicles. Big Bertha's punch pushed a large truck completely across the wharf.

Incredible as Big Bertha's effectiveness proved that afternoon, the hero of the day was Senior Operator Joseph V. "Rocky" Roquemore, alone in *Fireboat No. 3.* The 16-year-old wooden boat was not much of a firefighter. Only 38-foot-long and driven by a 275 horsepower gasoline engine, its pump could put out but 290 gpm.

Spotting many sailors and workers who had been blown or jumped from the dock and landing craft, Roquemore gave full throttle and steered the old boat close to the radiant heat and residual fires around and under the wharf. Tossing life preservers to the men in the water, he began pulling survivors on board. Many were badly burned or suffered other injuries.

Quickly putting them ashore where medical aid awaited, Roquemore sped back to rescue more sailors and workers in imminent danger of drowning. He was credited with saving 17 lives. For lack of safety precautions aboard the *Fredericksburg,* 11 sailors and five workers died. Thirty-five others were injured.

Roquemore's feat remains unparalleled in the history of the Los Angeles fireboat fleet. Although the Los Angeles Fire Department has medals for such bravery, Roquemore's heroism was never officially recognized.

Fire Chief John H. Alderson explained: "Rocky was only doing what firemen are paid to do and what duty compels them to do."

Fireman Joseph V. "Rocky" Roquemore, alone aboard Los Angeles' *Fireboat No. 3*, single-handedly rescued 17 sailors and workers who had jumped or were thrown into the water when explosions and fires erupted from Navy landing craft and a wooden dock on October 21, 1944. The $10,000 boat, commissioned March 13, 1928, was wooden and gasoline-powered; further adding to the dangers Roquemore faced during his rescues which remain unparalleled in the history of the Los Angeles fireboat fleet.

From the Collection of Bill Dahlquist

National security often caused authorities to withhold details of wartime fire, especially those in dense industries. Few details were released about this May 30, 1945, fire among U.S. Navy warships at Los Angeles. Shipyard and drydock tugs, along with Los Angeles Fireboat No. 2 joined in the battle.

From the Collection of Bill Dahlquist

During the closing days of the war the Empire State Building was struck at 9:52 a.m., Saturday, July 28, 1945, by an Army Air Force bomber that smacked the 78th floor level after its pilot became lost in fog blanketing Manhattan. A four-alarm fire, fueled by high octane aviation gasoline, spread through elevator shafts in what was, at that time, the tallest building in the world. Fourteen persons were killed.

In Japan, meanwhile, Tokyo's fire chief assured Emperor Hirohito that his department was completely prepared for American air raids. The Imperial Palace was safe, too. Tokyo had 287 fire stations and a fleet of 1117 pieces of apparatus covering an area of 213 square miles and protecting more than seven million people. The city had boosted its prewar fire personnel from 2000 to 8100 firefighters, including 2700 teenaged youths who were given junior fire department badges and assigned as fire alarm operators and messengers.

That was the good news for the Emperor. He would shortly learn the bad news.

Starting March 9, 1945, General "Hap" Arnold's 20th Army Air Force bombers began pounding Tokyo with an eye-for-an-eye vengeance that sparked the rallying cry, "Remember Pearl Harbor." The bombers came in waves of two or three planes at a time and from varying, crisscrossing directions.

Great chunks of Tokyo quickly blazed. Balmy four-mile-an-hour winds in the city rapidly grew to hurricane velocities as seemingly neverending clusters of incendiaries ate whole areas. These fires merged with others.

Pillars of smoke and flames, jetting thousands of feet, turned from orange to blue as heat ignited combustible gases inside the plumes. Condensing they showered Tokyo with black rain.

Fiercely violent thermal updrafts at 6000 feet flipped B-29 bombers upside down and forced crews to grab oxygen masks. For pilots to maintain altitudes against these updrafts was equivalent to diving at speeds exceeding the bombers' design capabilities. A B-29 became engulfed in a cyclonic updraft. The bomber did a 360-degree loop resulting in forces that slammed the captain

Some 65 Japanese cities burned following incendiary raids by Army Air Force bombers. Hiroshima and Nagasaki are at the southern tip of Japan. Tokyo is midway along the eastern coast.

and crew against the ceiling and plummeted the plane into a 12,000 foot freefall. The pilot managed to pull out just before hitting the ground.

Bombs hit precisely on target. Firemen fled from their apparatus and stations for air raid shelters. How could any of them fight fires when the blast furnace-like hurricane winds knocked them down? When they did manage to get to their feet they were flattened again by flying trees, parts of automobiles and debris.

Five hundred pound clusters of incendiary bombs from two Air Force bombers caused heavy damage, March 16, 1945, to Kobe's waterfront district. The photo was taken from a B-29 at an altitude of 15,000 feet. Note earlier damage (lower left) caused by another raid.

Tokyo's 559 motorized pumpers of 450-gallons-per-minute capacity inadequately protected the city. Firemen had a few 1920 vintage American-LaFrance 750-gallons-per-minute pumpers which had to be hand-cranked to start. As in the rest of Japan, most fire engines were drawn by hand and pumped by hand. Hose was rotten, leaky and of patched linen. It burst at relatively low pressures exceeding 100 pounds-per-square-inch.

Four miles of Osaka's industrialized waterfront and a large portion of the city beyond it were pummeled in this June, 1945, attack.

Tokyo's blazing, smoke-shrouded waterfront was euphemistically protected by three tiny fireboats. Each put out but 500 gallons-per-minute from pathetically small-caliber nozzles. They were the largest in all of Japan but only slightly larger than nozzles used by American firemen to douse grass fires.

Firemen knew but one firefighting evolution: Arriving at air raid fires they unloaded a reel of hose from the engine, stretched it for 131 feet, connected the hose suction to the pumper, then laid two sections of hose back to the pumper. As they learned during prewar training, this procedure required seven men. Less than that number of available firemen, the rig was taken out of service.

Incendiary bombs took out Tokyo's business district. Fire-resistant buildings (background) withstood the heat, but fires entered the unprotected windows of other buildings which were destroyed.

All Japanese fire companies had individually distinctive standards like this one at the front of the fire engine. Firemen took these standards to fires and placed them where they declared they would make a stand and not let the flames pass them by. This 1920 vintage American-LaFrance 750 gallons-per-minute pumper was one of a few in Tokyo.

Another firefighting tradition that outlived its usefulness that day, were distinctly marked banners mounted outside firehouse doors. These flags were trotted to fires by standard bearers and planted at a point where firemen vowed they would not let flames pass. Hundreds of banners were consumed.

High-explosive bombs knocked out water supplies, telephones, and fire alarm box systems. Street corner-mounted fireboxes had long fallen into disrepair. More than 90 percent of Tokyo's fires were spotted from watchtowers. Upon sighting a fire and guessing its approximate location, junior firemen pedaled their bicycles to the nearest firehouse. Nearly all were empty that day. The alarms went unanswered. Bombardiers easily picked out these towers and firehouses, then lined them up in their bombsights and pressed triggers.

Tokyo's water supplies were abundant, although mostly stored in barrels and concrete tanks holding around 70 gallons. There were but five automatic fire sprinkler systems in all of Tokyo. Three were in areas ignored by Air Force bombers. The other two failed for lack of water pressure to supply them.

Fire officials failed to consider fuel needs for apparatus during these prolonged fires. Running out of gasoline, Tokyo firemen left their engines, returned to their stations for more and raced back with fuel cans on their shoulders. They quickly became lost in the thick smoke. The apparatus was abandoned. Outside of Tokyo, firemen with 60 pumpers raced to aid Tokyo. Only two got there. The rest ran out of gas.

The entire shebang was under the leadership of police officials, another tradition dating to feudal times when knights guarded castles. Police directed firefighting as they would traffic. They closed off streets and ordered firemen to find another way into the fires. Only fleeing

Fire alarm boxes (left) like these in Osaka were unreliable and rarely used to turn in alarms. Around 95 percent of Tokyo's fires were spotted from watch towers such as this bombed-out tower.

Tokyo ablaze following the March 10, 1945, Twentieth Army Air Force raid. Police officers closed off streets to firemen so survivors could readily escape. Denied immediate access, firemen were delayed before they could mount an attack.

residents were permitted to use these streets. Firemen fumed over what was worse: Fires or the befuddled policemen who tried to direct their efforts.

Tokyo's fire chief was ready to reach for his hara-kiri sword when he stood before Emperor Hirohito after the raid. Nearly 84,000 were killed. About half of the city was burning. Among the dead were 125 firemen. Lost, too, were 100 firehouses and 186 fire engines. The chief could no longer guarantee the safety of the Imperial Palace. On that score he was right. B-29 bombers later pelted the palace.

From their bases in the Mariana Islands, the Twentieth Air Force focused upon a two-mile coastal strip extending from Tokyo south where more than half of Japan's military production was centered. Osaka, the Chicago of Japan, was battered. More than eight square miles inside the city were destroyed, including 48 fire engines which sputtered out of fuel and were abandoned to the flames. More than 68,000 buildings in industrial Kobe were destroyed and 116 fire engines burned beyond repair by four more doses of bombs. Japanese police chiefs desperately confiscated fire engines in villages and towns and sent them to fight fires in 65 cities. These smaller communities were left to protect themselves with water bucket brigades.

Police later were understandably reluctant to cooperate with damage assessment teams. A fire buff was not. This young man, Dr. Honde, knew more about Japanese fire departments than the chiefs. Dr. Honde had spent 10 years in the United States and England where he marveled at the expertise of professional firefighters while he chased fire engines. Along the way he earned a doctorate in chemistry from Ohio's Oberlin College.

Dr. Honde's ambition was to earn tenure at a Japanese university, but the war scotched that idea. His knowledge of Japanese fire departments and firefighting, together with his language fluency, made him "of tremendous assistance," according to the U.S. Strategic Bomb Survey teams.

Altogether, firemen in 65 cities suffered similar fates as their Tokyo colleagues did from 93,000 tons of incendiaries and 650 tons of high-explosives. When the B-25s were finished, nearly all of Tokyo lay in ruins. Only two more cities remained on the hit list.

Hiroshima on August 6, 1945, was one of Japan's major military centers. Air raid sirens sounded an alert at 7:30 a.m., but it proved to be another false alarm of many in the city of more than 340,000. People went about their routine business.

At 8:15 a.m., 45 minutes after the all-clear had sounded, the history of mankind entered a new era when the first atomic bomb exploded over Hiroshima. More

This hand-drawn fire engine was typical of most apparatus in Japanese fire departments. The pump was gasoline-operated but produced no more than 120 gallons-per-minute.

Hiroshima's Magirus-built aerial ladder never saw service as the atomic bomb wiped out this fire station, apparatus and the firemen inside.

Japanese fire brigades used hand-drawn fire engines which were the first line of defense when Air Force bombers dropped incendiaries. The pump was gasoline powered.

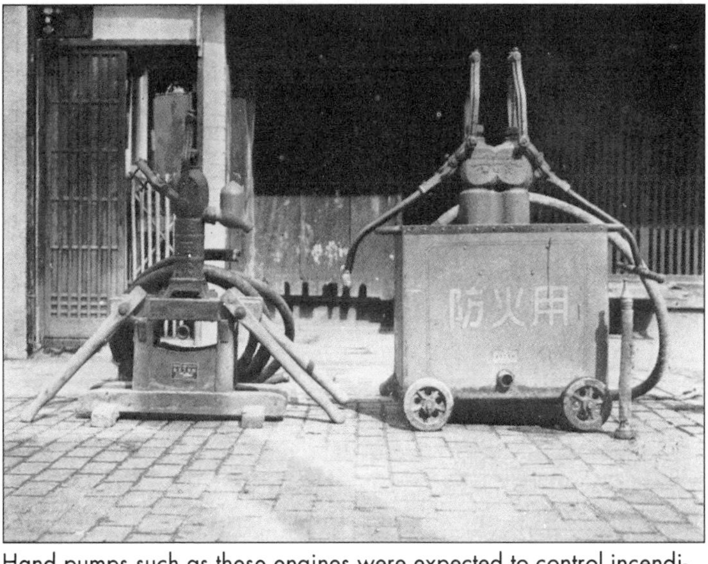

Hand pumps such as these engines were expected to control incendiary raid fires in Japanese cities. Motorized apparatus was virtually non-existent.

Hiroshima's fire apparatus was destroyed by the atom bomb dropped on August 6, 1945.

than four square miles of the city was laid barren and up to 80,000 people were killed, including every Hiroshima fireman. Their 16 rigs were destroyed.

A similar scenario played out three days later in Nagasaki. Following air raid alerts which proved to be false that morning, citizens went about their business, although they had heard rumors of a mysterious bomb explosion in Hiroshima despite censors' efforts to stifle the news. At 11 a.m., two B-29 bombers were spotted at high altitudes, but sirens remained silent. Two minutes later the second atomic bomb was dropped. The Nagasaki Prefecture Report said:

"An intense flash was observed first, as though a large amount of magnesium had been ignited, and the scene grew hazy with white smoke ... A tremendous roaring sound was heard and a crushing blast wave and intense heat were felt," up to 40,000 were killed.

World War II ended nine days after Hiroshima and Nagasaki when Japan unconditionally surrendered. Especially regarding Hiroshima and Nagasaki, the Japanese people seemed as bitter with their government as they were with the American raiders. A common reaction: Since it was war, it was just Shikata-ga-nai "too bad."

The U.S. Strategic Bombing Survey later found that: "The real decision to end the war came just after the greatest (March 9, 1945) attack upon Tokyo ... and the demonstration of fire devastation immediately following fire attacks on Nagoya, Osaka and Kobe ... After these great fires the people, as certainly as the government, could clearly see that the war was lost."

The same survey of the results of bombs dropped on Germany said:

"If the German people had, as a result of the raids, been at liberty to vote themselves out of the war, they would have done so before the final surrender."

In further support of these findings, U.S. Army Major Forrest J. Sanborn told the 1946 annual meeting of the National Fire Protection Association:

"If one needs to be reminded that civilians were the real sufferers, he need only to recall that the number of Japanese casualties in the Japanese homeland, inflicted entirely by our Air Force during a six month period, was twice the Japanese military casualties inflicted by our combined Army, Navy, Air Force and Marines during a 45-month period...

"The effective use of fire as a major offensive weapon in war has been shown to the armed forces of every country in the world. If we are to make the most of the offensive effort of our armed forces in any future war, which God forbid we must, much more effective use of the knowledge and ingenuity of fire prevention and fire protection engineers must be made by the scientific and military authorities responsible for the winning of it."

Although World War II was over, there still remained the secret story of a catastrophe in Bombay. Not until many years later was that story told.

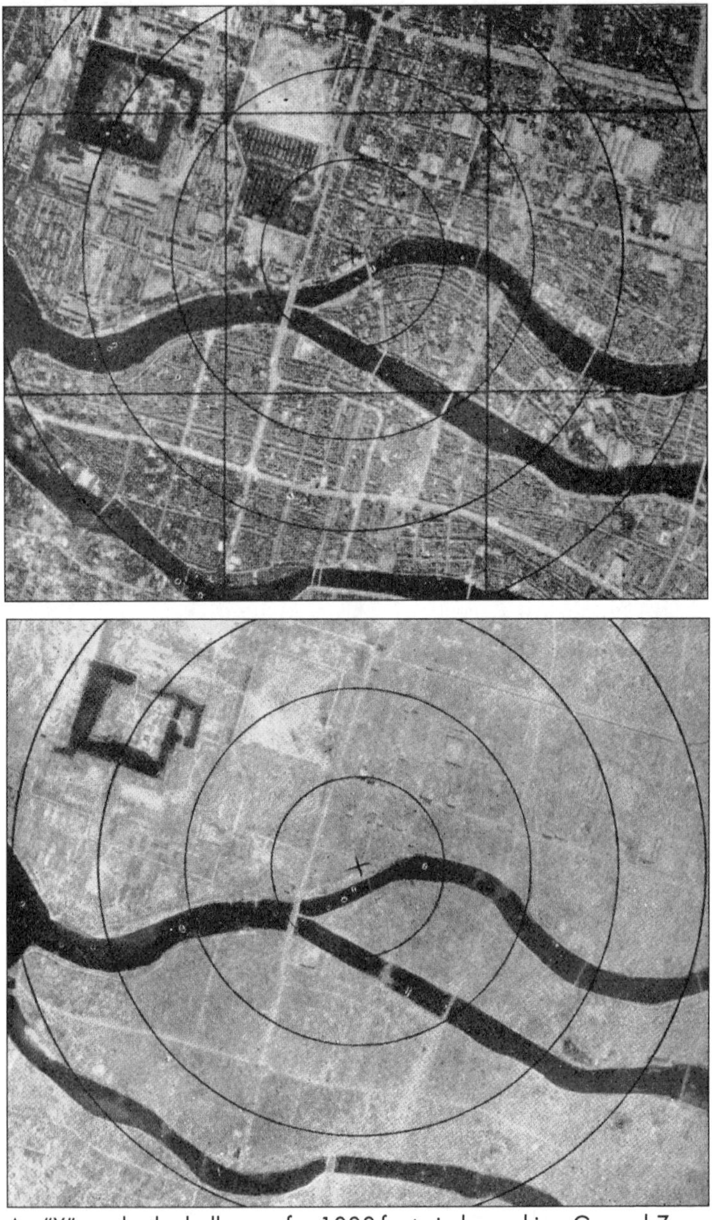

An "X" marks the bullseye of a 1000-foot circle marking Ground Zero immediately preceding the August 6, 1945, dropping of an atomic bomb on Hiroshima. The other photo shows Hiroshima after the bomb detonated.

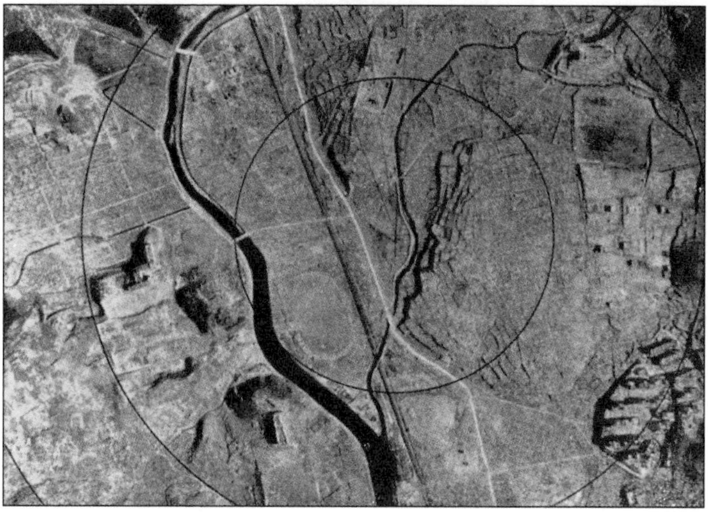

United States Army Air Force photos show Nagasaki on the day of the August 9, 1945, attack and the obliteration of the city following the bomb drop.

CHAPTER V

The Best-Kept Secret Of World War II

Bombay had been tinder dry for nine months and fairly begged for a spark to touch off a conflagration. Warm breezes from off the Arabian Sea only seemed to make the 88-degree heat more oppressive at high noon on that Friday, April 14, 1944, in the lobster-claw-shaped city hanging at the tip of tiny Salsette Island on the west central coast of India.

The wind, teaming with the dangerously low humidity, could whip up a full-blown catastrophe if a fire got started in the six-story tenement district or in the crowded dock area along the outer edge of the eastern pincer. It was there that godowns (warehouses) and freighters were loaded with explosives, war supplies, food and cotton, Bombay's major industry. Norman Coombs, chief officer of the Bombay Fire Brigade, would breathe a lot easier when the June monsoons soaked the sun-parched city.

The Gateway to India had never suffered a catastrophic fire — but Coombs had a feeling that his luck could not last indefinitely, especially with dangerous cargoes streaming into the docks. Since the war started he had fought 60 shipboard fires, 15 of them involving ammunition cargoes. If one of these ships ever blew up the city would be sprayed with red hot debris, touching off countless fires.

A U.S. Coast Guard photograph of the *S.S. Fort Stikine* as she appeared before the explosion.

On Tuesday, two days before that fateful afternoon, Coombs had heard that an ammo ship was burning at Berth 1 in crowded Victoria Dock. Coombs investigated and found the *SS Fort Stikine*, operated by the British Ministry of War Transport, freshly arrived from Karachi on the final leg of a five week voyage that brought it from Birkenhead, England.

The 7143-ton freighter seemed shipshape — but buried deep in her cargo of 8700 bales of Punjab cotton from Karachi, lube oil, fish, fertilizer, scrap and 1400 tons of ammunition were the ingredients for one of the most devastating non-nuclear explosions of all time, a disaster that was to become one of the best-kept secrets of World War II. By comparison, the Texas City affair was a dud. Not even the famous ammunition ship blast in Halifax, Nova Scotia, December 6, 1917, could match it.

Most fire chiefs are pessimists, and Coombs was no exception. He was totally against the clause in Defense of India Rule No. 88 permitting the Bombay Port Trust's military officer to waive the regulation forbidding ships with dangerous cargoes from entering the harbor. The military officer had argued that war needs demanded rapid turnovers in shipping space. Time did not permit the lightering of ammunition from anchorages outside the port of dock side cargo sheds, the officer had said.

As he always did when ammo ships were berthed in Bombay, Coombs ordered a fire engine and crew to stand alongside the *Stikine* until the explosives were off-loaded. "If you see any sign of fire," he told the pumper's section leader, "turn in a No. 2 alarm." Coombs went home satisfied there was no fire aboard the *Stikine* and passed off the report as a rumor, one of hundreds circulating through the city.

The rumor persisted, however, as did another one: the *Fort Stikine* carried nearly $6 million in solid gold bars. This report was perhaps more easily nailed down. The ship had been built two years earlier in British Columbia, and took her name from a gold mining settlement formed along the Stikine River during the 1862 Gold Rush.

As Coombs settled back at 1 o'clock that lazy Friday afternoon he could not escape a two-pronged, ever-present worry: the weight of the war rested upon Bombay, and the safety of its 1,500,000 natives and thousands of British Indian and American servicemen rested upon his shoulders.

Bombay had no fireboats to protect her 104 acres of docks.

The seasoned 46-year-old firefighter knew his small 12-station brigade of 156 officers and men could never stop a major conflagration with only 72 pieces of apparatus, much of it antiquated. Bombay had no fireboats to protect her 104 acres of docks. Coombs would shortly feel this serious flaw in the city's fire protection setup.

A few of Coombs' rigs were relatively new 1000 gallon pumpers. He concentrated them in the congested high-value district around the docks. The rest were too old or had such low pumping capacities that they were useless. Many Bombay factory, textile mill and oil tank farm owners were so fearful for the safety of their plants that they maintained their own fire engines and employed full time crews to operate them.

Bombay was the British Empire's third largest city, a strategic link in the flow of men and supplies to the Far

East and to the Russian army via the Persian Gulf—and No. 10 Downing Street had even more vitally important plans for its post facilities. Although Winston Churchill would not mention it publicly until he wrote his war memoirs, London and Bombay insiders called the top secret scheme "Churchill's Ace-in-The-Hole." They knew he planned to make Bombay the springboard for an Allied invasion of Europe. Churchill was positive this new format would save Eastern Europe from postwar domination by the Soviets. The only fly in the ointment was Turkish Prime Minister Ismet Inonu's hesitancy to enter the war on the Allied side.

Coombs knew that thousands of troops were poised in Bombay and huge quantities of ammunition and supplies were being stockpiled in godowns. The *Queen Mary*, converted to troop carrying, was a familiar sight in Bombay harbor. As soon as Inonu could be won over, the invasion force would head up the Persian Gulf, then move through Syria and Lebanon into Turkey.

Churchill was worried that Hitler or Hirohito might figure out this strategy and blitz Bombay from the air. If the city was lost and its three major docks—Victoria, Prince's and the newer and larger Alexandra—were knocked out of action, the Allies would have a hard time keeping the supplies flowing to the new front.

At 1:30 that afternoon when the dock whistle called long-shoremen back from lunch, Mahomed Taqi and his gang of stevedores went back down into the *Fort Stikine's* No. 2 forward hold. Twenty minutes later Taqi noticed wisps of smoke curling up from between the bales of cotton on which he was standing.

"Fire!" yelled Taqi. The hatch boss ran for the *Stikine's* fire alarm while cursing the longshore gangs for smoking in restricted areas. Crewmen were dragging hose across the deck as Taqi and his men climbed from the hold.

Coombs' men on the quay heard the alarm, grabbed a reel of hose and hurried up the gangplank with it. The section leader leaned over the hatch coaming and peered down through the smoke. He figured he'd look ridiculous if he pulled a No. 2 alarm for something as insignificant as this.

Stevedores were now coming up out of the other holds to see what was causing all the commotion. "It's only a little fire!" the hatch boss shouted. "Go back to work. And you, Taqi, take your gang and bear a hand over in Number Four." Winches whined again and the doomed stevedores continued the unloading.

It was now 1:10 p.m.— the fire had been discovered 20 minutes ago. The brigade section leader noticed that the hoses were not putting out the blaze. The smoke was getting thicker and the heat worse. He told his sub-leader to find a telephone and call for a No. 2 alarm.

Confusion now began to take hold of everyone clustered around the coaming. Nobody, fire brigade men or ship's officers, thought to batten down the open lower hatchway and turn on the *Fort Stikine's* steam injectors. Live steam jetting into the cotton would have checked the blaze until some better plan of attack was devised.

The sub-leader ran to a telephone kiosk, intent upon dialing 2-9-0 to reach the brigade Central Control. He became flustered when he found the phone had no dial. Far down the quay, 170 yards away, was a fire alarm box. He raced for it.

The clock in Bombay's ornate Victoria Railway Terminal, 1000 yards southwest of the *Stikine*, read

2:16 p.m. when the sub-leader pulled the firebox hook down. He returned to the ship without waiting to direct incoming apparatus to the fire.

The signal was received on the register of the Alexandra Dock Fire Station half a mile south of the *Stikine*, and two large pumpers clanged from the station. The alarm was simultaneously relayed to the brigade control room where dispatchers regarded it as an ordinary alarm such as any dockworker might turn in. Certainly, this call did not rate a heavy No. 2 alarm response of men and equipment.

Eight minutes later, at 2:24 p.m., the serious leader aboard the *Stikine* looked up, puzzled, as the distant clamor of approaching fire engines ceased. He leaned over the ship's railing and looked down the quay. Instead of the 10 pumpers, turntable ladders and emergency tenders he had expected, only two rigs pulled up to the distant firebox.

The section leader waved and shouted to attract their attention, but the Alexandra Station firemen did not see him. Just as they were about to write the call off as a false alarm and return to their quarters, Station Officer Mobarek Singh saw a misty cloud of smoke spiraling lazily from the *Stikine*. Knowing explosives were aboard the ship, he immediately sent in a second alarm. The time 2:30 p.m. There had been a delay of 40 precious minutes since the fire was discovered. It was a handicap that would prove fatal.

The instant the No. 2 message was recorded in the control room the chief dispatcher plugged in a line to Coombs' home. Coombs did not take time to change from his tropical weight linen jacket and slacks into his boots and firecoat, a hasty decision he would shortly regret. Eleven minutes later, after a wild, clamorous ride through crowded and narrow streets, Coombs' driver, Mahomed Siddick, wheeled up alongside the *Stikine*.

Coombs hurried up the gangplank and stepped over the snarled hoselines snaking along the deck. He squinted into the murky hold, rubbed his eyes and coughed. "Damn this cotton! It gives out a worse bellyful of smoke than any other kind of a fire."

Coombs often said he would sooner fight a 4-alarm warehouse blaze than a few bales of smoldering cotton. The brigade had fought lots of cotton fires—three in recent weeks. The bales took days to unload, rip apart, and soak until water penetrated the fibers.

"How many lines have you got working?"

"Thirty-two, sir," said Assistant Chief Harry Palmer, "counting those operated by the *Stikine's* crew."

"We better go below and see if we can find the seat of the fire," said Coombs. Singh and Fireman Arthur Reynolds put on smoke masks and were lowered by ropes into the charcoal gray smoke.

From the *Fort Stikine's* captain, A. H. Naismith, Coombs obtained a copy of the ship's manifest. It showed a thick carpet of ammunition spreading across the bottom of the hold below the cotton bales and the 6000 cubic feet of lumber and a three-ton chunk of scrap iron. Above this cargo and stowed in the 'tween decks clear up to the underside of the main deck was another heavy layer of ammunition.

"Good Lord!" stormed Coombs. "There's 370 tons of TNT directly under us!"

Anxiously, Coombs questioned Naismith—and learned that, although the *Stikine* had been in port for two days, only a few crates of explosives had been unloaded.

"Good Lord!" stormed Coombs. "There's 370 tons of TNT directly under us! Why in God's name haven't you unloaded this stuff?"

"They told me after I dropped anchor that there would be a delay before special lighters and vans became available to get those pallets of fish off. That confounded stink was driving us crazy in this heat."

Singh and Reynolds tugged at their lifelines and were hauled up.

"Sorry, sir," said Reynolds, slipping off his mask. "We can't work down there. It's too bloody hot even with the hoselines wetting us down and the smoke's so thick you can't see your hand in front of your face."

"But you've got to go back down there," insisted Naismith. "You've got to get the gold out."

"Gold?"

"There's 124 ingots in the specie locker!"

"Where is it? We'll try to get it up."

"You can't. The locker door is blocked by all that cotton!"

"What the hell is gold doing down there?"

"The Bank of England is sending it to the Reserve Bank of India to help stabilize the rupee."

Coombs leaned over the hatch coaming, and scowled when he felt the heat, more intense now, gushing up from below. How was he ever going to find the gold buried somewhere in that mess?

"We'll flood the hold," Coombs told Palmer. "If that fails we'll have to scuttle. I hate to think of sending all that gold to Davy Jones, but we can't wait for the heat to set off the ammunition."

It did not occur to Coombs to make certain the steam injectors were working. Surely, he thought, that would have been the first action of the *Fort Stikine's* officers when the fire was discovered.

Captain B. T. Oberst, an Army ordnance expert, came aboard when he saw the smoke and studied the cargo chart. His eyes opened wide. "You've got how much ammo aboard?"

"Close to 1400 tons," calculated Naismith.

"My God!" gasped Oberst. "There's enough TNT under our feet to pack the wallop of 150 blockbusters exploding at one time! Scuttle immediately. If you don't you'll blow up the whole bloomin' harbor and the city with it!"

At the nearby Burmah-Shell Oil Company tank farm, Fire Officer Leslie Hall was completing arrangements for a dinner party at his home Saturday evening. He wanted to make sure that two of his friends, Firemen Reynolds and Robert Andrews, had not forgotten about it. When Hall called the brigade he was told they were at Victoria Dock; curious, he climbed into his station wagon and drove to the dock. He found them leaning over the *Stikine's* rail.

"What's burning?" Hall called up to the pair.

"Just a little fire in some cotton," said Reynolds. "It looks worse than it really is. We'll probably have it out by tea time."

After reminding his friends about the party Hall returned to the tank farm. He never saw Reynolds or Andrews alive again.

It was now 2:45 p.m., one hour and 55 minutes after the fire had been discovered. Fireman H. V. Dayaram, driver of one of the fire engines lining the quay, was adjusting the pressure on his pump when he happened to notice a patch of blistering paint on the port side of the *Stikine's* hull at the aft end of No. 2 hold and at a level below the 'tween decks.

"Chief Coombs!" yelled Dayaram. "Come quickly!" Coombs ran down the gangplank and looked up to where Dayaram pointed. Although the patch was 25 feet above him he could feel the heat. No wonder his 32 hoselines were not boring into the seat of the fire: the patch showed that it was centered at a point inaccessible from the deck.

Coombs ordered his men to rig a platform beside the patch so Assistant Chief Palmer, could cut a hole through the *Fort Stikine's* three-eighths-inch-thick hull with the brigade's portable acetylene torch. Coombs hated to lose the quarter hour necessary to do the job. Did he have time? Coombs knew Palmer would take a brutal beating from the blister, which was heated to at least 2000 degrees, judging by the glow. Once he cut through the hull Palmer would have to work blind in the suffocating smoke spewing from the slit.

The platform was completed shortly after 3 o'clock and the firemen passed the acetylene cutter up to the assistant chief. Palmer tried to ignite the torch. It spat only black smoke. He flicked the switch again and again. The torch refused to work.

"Send for another one," snapped Coombs. A field security sergeant telephoned for a replacement. At 3:08 p.m., a deputy docks manager cancelled the request. He did not live to explain why.

Captain Naismith was surrounded by a group of port, military, naval, salvage and fire officers, including Co. J. R. Sadler, general manager of the docks and railways section of the Bombay Port Trust. "There's no question about it," said Sadler. "Regardless of the gold, you've got to scuttle before that ammo blows us sky high."

"Yes," agreed Coombs and Marine Surveyor C. W. Stevens.

"No!" said Naismith and the *Stikine's* chief officer, W. D. Henderson. "Let's try a bit longer to save her. Besides, we can't scuttle until Lloyd's of London is notified."

"We can't wait!" insisted Sadler.

The *Stikine* would have to be taken into deep water outside the harbor to be scuttled, for she drew 20 feet aft, and the depth of the tidal basin was only 26 feet, barely enough to cover her double bottoms. After the ship had left the dock Coombs would have to gamble upon the inadequate fire pumps of the *Stikine* to hold the fire in check until they reached open sea. If only the government had provided him with fireboats!

It was now 3:15 p.m. The clock was running out on Bombay. Naismith made no move to call for tugs to stand by to tow the *Stikine* out of the harbor. He did not

want to be responsible for paying for their services if they were not used.

The great debate continued as the fire burned closer to the explosives. Nobody thought to follow port rules requiring that Commodore Rattray of the Royal Indian Navy and Captain Curtis, naval-officer-in-charge, Bombay, be notified of all shipboard fires. They had supreme power to overrule Naismith and scuttle the *Stikine* immediately. Neither of them had the faintest knowledge of the catastrophe that was building up while they sat blissfully at their desks in Bombay's Government Secretariat 3000 yards to the southwest.

Nobody bothered, either, to notify the Commissioner of Police of the gravity of the situation. Police could have cleared the area of swarms of workers, hundreds of whom were about to die needlessly.

Coombs looked at his watch, 3:30 p.m. What was holding up the new torch? He sent an aide to find out. Coombs felt the *Stikine* listing to starboard—the thousands of gallons of water being pumped into her hold had thrown the freighter off balance.

At 3:45 the heat blister had spread like an infection across the hull. It glowed cherry red. Steam rose from the water for 20 feet around the *Stikine*. Her thick iron deck plates turned unbearably hot. Coombs helped his men lay planks across the deck to keep the hoses from burning. The planks smoldered. A fireman doused them.

Coombs peered over the hatch coaming. He could tell from the terrific heat and boiling smoke how much pressure was building up. Too, the smoke was changing to a yellowish tint, a sign of a badly bottled-up fire.

Shimmering waves of heat rose from the deck as the tropic sun beat down upon the *Stikine*. Sweat soaked Coombs' linen suit. The thick smoke made his eyes smart. Tears streaked his grime-caked face. His nose streamed. Where was that torch? The low rumbling roar of the fire below decks drowned the steady pumping of the long line of fire engines on the quay and the hiss of water spewing from the nozzles.

Through a gap in the smoke Coombs saw the three-ton chunk of scrap iron glowing red hot. What he could not see were the bales of burning cotton floating on the slowly rising water. Coombs was worried that the fire would burn *down* to the carpet of ammunition. He didn't imagine the fire might float *up* to the TNT stowed in the 'tween decks.

The stevedores came up from the other holds to watch the excitement. Everyone was too preoccupied with No. 2 to notice that No. 4 hold—containing twice as much TNT—was not battened down. Stray sparks were floating into the open hold.

The *Stikine* listed more sharply, and a knot of terror tightened in Coombs' stomach. Even if Captain Naismith agreed to scuttling, the water sloshing in the ship's belly made her too unstable to head for sea.

It was now 3:52 p.m. The *Fort Stikine* had been burning for more than two hours. Despite the danger, no attempt was made to warn the 39 freighters in Victoria or the adjacent Prince's Dock to the north or Alexandra Dock, half a mile south, below narrow Carnac Bunder and the Rice Market. All three basins were filled to capacity. Among the vessels anchored in Alexandra were three ammunition ships, an oil tanker, and an American transport, the *SS Mariposa*, loading 4000

United States troops for Australia and scheduled to depart at high tide.

Masters of ships berthed near the *Stikine* were not worried. They had all seen these bothersome cotton fires and felt safe so long as the *Stikine* did not fly the mandatory International Code Signal red flag indicating a dangerous cargo. There was to be considerable speculation later on the probable connection between the *Stikine's* lapse and her cargo of gold. Many believe the harbor rule was deliberately flouted because nobody wanted to advertise the floating treasure chest. It has also been suggested that the failure to unload the ammunition sooner was due to the government's desire to reach the gold.

Fifty yards astern of the *Stikine* at Berth 2 lay the Dutch freighter, *SS Jalapadma*. Her crewmen soon tired of watching the firemen and ship's officers who seemed to be doing little but argue and pump more water into the smoking *Stikine*. Not much of a show, they agreed—and went below to look for other entertainment. Most of them never saw daylight again.

At the Red and Yellow Gate entrance to Victoria Dock policemen were checking passes of workers coming on duty or going home. "Hurry up!" said a stevedore who was in the long outgoing line at Yellow Gate. "Don't you know there's an ammo ship burning over there? She's liable to blow!" The policeman, immune by experience to waterfront rumors, ignored him.

The only explosion Bombay might reasonably have feared that afternoon was from the hatreds smoldering between the natives and the British. Tensions had flared into rioting during recent weeks. Nehru and

Gandhi were in prison and Bombay's mayor was serving an 18-month jail term for helping to incite violence.

The tower clock in Victoria Station hurried toward 4 o'clock, tea time in Bombay. Leslie Hall ordered a cup brought into his office. A mile west of the docks, D. C. Motliwala, a retired Parsee civil engineer, was sipping iced tea in the shade of his third floor verandah.

The cares of the war seemed distant, too, in the business district of Bombay a few miles south and west of the docks. The Taj Mahal Hotel bar was crowded. In the teeming bazaars American soldiers and British tommies mingled with bronzed natives and scrofulous beggars while haggling over the prices of teak carvings, joss sticks, ivory elephants and saris. Tonight the troops would fight off their boredom by taking in a movie—or so they thought. Bombay's cinemas were showing *The Girls He Left Behind*, starring Carmen Miranda, and *Sahara*, with Humphrey Bogart. The humidity had lulled nearly everyone in Bombay into a relaxed feeling that all was well.

Coombs wished he could share their optimism, for at that moment the smoke billowing from the *Stikine's* hold suddenly turned jet black. A streak of flame shot as high as the foremast.

"She's into the ammo!" gasped Coombs as a solid ball of fire swished straight up like a roman candle.

"Get off the ship!" shouted Coombs. "Get off the ship! She's going to blow!"

Assistant Chief Palmer and most of the firemen fled down the gangplank. Siddick sprinted for Coombs' car and drove wildly down the quay, careening past the fire engines whose drivers stood and gaped in disbelief at the volcanic flame spewing from the *Stikine*. Then they dove for shelter underneath their rigs.

Coombs kicked off his shoes, leaped into the water and swam for a towboat. He braced himself for a mighty blast. But none came.

Masters of other freighters in Victoria wondered what was making burning cotton do such crazy things. They looked again for the red flag. Two airplanes passed low over the docks and headed in toward Bombay.

The towboat picked up Coombs and put him on the quay where he called his men together and set them to work fighting fires started by sparks falling among cotton bales stacked on the dock. In a few minutes, when the ammunition stopped popping, they could resume the battle aboard the *Stikine*.

The thundering boom rocked Bombay and stopped the clock in Victoria Station.

When Siddick reached Yellow Gate he remembered that Coombs had not come down the gangplank like Palmer and the other firemen. "The chief must be trapped!" thought Siddick. "I've got to save him!" He wheeled the car around and raced back down the quay. At 4:06 p.m., two hours and 16 minutes after the fire was discovered, the forward section of the *Fort Stikine* exploded with a shattering roar in Siddick's face.

The thundering boom rocked Bombay and stopped the clock in Victoria Station. One thousand miles away the seismograph at the Simla Meteorological Station in the foothills of the Himalaya Mountains registered an earthquake.

Hundreds of longshoremen and ships' crews were mowed down by the hail of white hot pieces of the *Stikine*, exploding machine gun bullets, anti-aircraft shells and shrapnel. Yellow Gate was a blood bath. The policeman who had been checking passes lay disemboweled. His hand still clutched a stevedore's identification card.

The force of the blast stripped Stevens naked and slammed him onto the dock. He sat up and looked around for Naismith and Henderson, who had been at his side when the *Stikine* blew up. Nobody saw them again. Colonel Sadler had also vanished. So had Singh and Palmer the brigade section leader who could not get excited over a little bit of smoke.

Retired Engineer Motliwala had just finished his tea when he heard a crash upstairs. A blob of brilliance streaked past his face. The thing landed with a thud at his feet. Motliwala stared dumfounded at the $41,000 solid gold ingot that had nearly hit him in the head.

Dozens of these 28-pound cakes of gold rained upon Bombay. They landed in the crowded bazaars, in gutters, on rooftops and crashed through windows. One pierced the hood of an automobile. Several natives were knocked unconscious. Beggers, servicemen and merchants, awed by the ungodly explosion they had heard, looked up to the heavens, then fell on their hands and knees to scramble, claw and fight for the gold. Beggars became rich men as they scooped up their treasure, stuffed it into their tattered robes, threw away their crutches and ran to their hovels to examine and caress their fortunes, Hindu and Muslem natives knelt and faced the sky while crying their hosannas for celestial favors.

Leslie Hall nearly upset his tea cup when his office shook from the concussion. He was sure an oil tank had blown up. He rushed into the yard, saw the cloud of black smoke rising from Victoria, and hurried to the docks to look for his friends.

Nurse Florence Ketchell, assistant matron at St. George's Hospital near Victoria Terminal, was pulling her white nylon uniform over her head preparatory to going on duty when she heard the blast. Opening her door she found that the door bolts were bent over like hairpins.

Miss Ketchell hurried to the casualty room to prepare for casualties which would soon start pouring in. Outside, she saw a bullock, still harnessed to an overturned cart, lying dead at the hospital gate. The animal had suffered no apparent wounds, but the concussion had collapsed its lungs.

Lt. Col. R. V. Morrison, superintendent of the hospital, sent out every one of his ambulances as reports flooded the switchboard that people were sprawled in the streets for dozens of blocks around Prince's and Victoria Docks. He quickly looked over the hospital's records and discharged every patient who could walk, telling them to return when the emergency was over.

Thousands of Bombayans figured the blast had something to do with the two planes flying over the city: at last their nightmares of air raids were being realized. A wave of panic rippled through Bombay. Hundreds grabbed their tin helmets and dashed for air raid shelters.

Rattray and Curtis sat bolt upright when they heard the blast and felt the Government Secretariat tremble. "I say!" said an aide. "That *was* a bad one!" Rattray stared from his office windows at the huge, ugly cloud of smoke billowing high over the docks.

Coombs had been facing away from the *Stikine* when she exploded. He was pitched across the quay and lay sprawled on the dock, too stunned to feel pain, although his back was burned and blood ran from a hundred cuts where debris had peppered him. He soon became aware of a strange breeziness: his trousers, except for the waistband and pockets that dangled in charred shreds, had been blown off. He tried to remember why he had not taken the time to put on his protective firecoat and boots before coming to the docks.

Coombs shook his head to clear it of the sharp ringing sensation and looked around. The devastation was incredible.

The forward section of the *Fort Stikine* was a mass of boiling flames and billowing black smoke. The blast had driven an enormous tidal wave across the basin. Hawser lines of ships had snapped as if they were kite strings. Nine mammoth freighters, free of their bonds, bobbed like plastic boats in a bathtub, smashing into each other and crushing smaller boats that got in their way. Tugs, water boats and dredges had ridden the crest of the wave and were slammed against the quay on the opposite bank. Some were sinking, others were aground.

Flaming debris spouting from the *Stikine* had set fire to 17 ships in Victoria and Prince's Docks. Four burning freighters were jammed together as they traded flames with each other while turning slowly in the basin. Twenty-four huge dock cranes lay in massive tangles of girders, booms and cables.

POSITION OF SHIPS BEFORE EXPLOSION

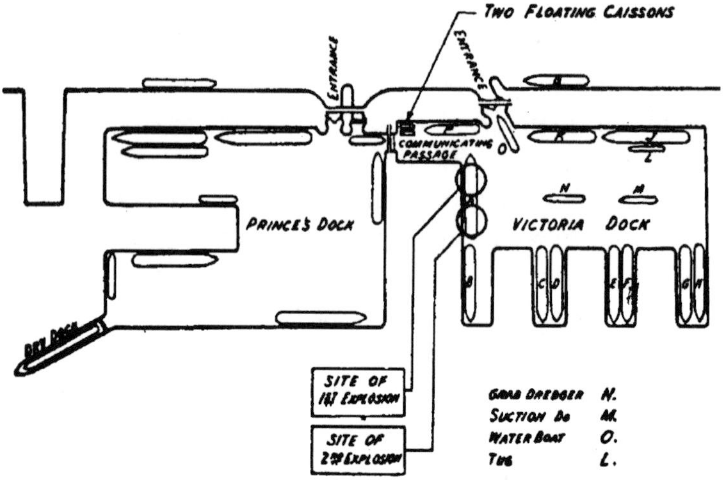

POSITION OF SHIPS AFTER EXPLOSION

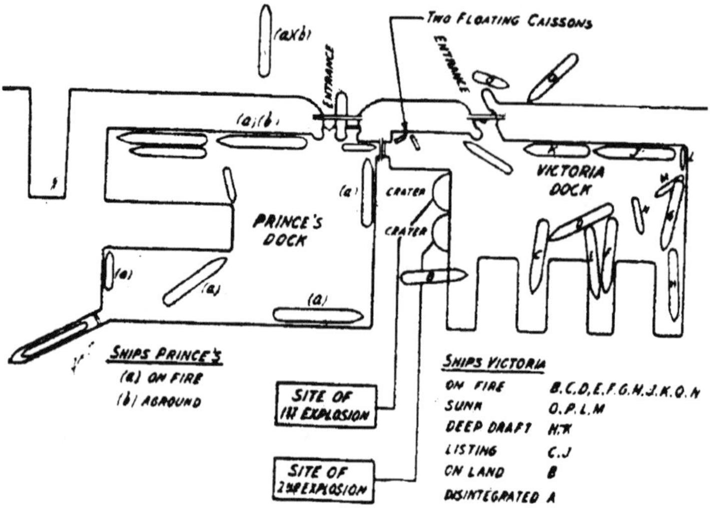

Blazing hunks of lumber and burning bales of cotton landing among the flattened godowns were kindling dozens of fires among the strewn contents of the warehouses. The wind was tossing firebrands through their shattered windows and fanning more fires among the stacks of crates inside.

Coombs had no way of knowing how many of his men were dead or injured. Corpses and chunks of bodies were strewn among the badly hurt firemen and dockworkers moaning on the quay. He wondered how he could save the waterfront or even Bombay itself without men and equipment; his brigade's heaviest duty rigs were now scattered, wrecked and burning, their tires stinking blobs of sizzling rubber and their hoses turning to ashes on the reels.

Coombs sent word to the brigade control room to dispatch the rest of the city's fire apparatus and to call back the off-duty platoon and order them to bring the reserve rigs to the docks. "Also tell them to fetch me another pair of trousers, a firecoat and some boots," he added.

Within minutes all of Bombay echoed with the shrieks of sirens as fire equipment raced down Carnac and Sandhurst Roads toward the mountain of smoke rising from the docks. Lt. Col. B. S. James, commander of 2000 troops in a transit camp, saw the billowing black cloud and sent lorries loaded with soldiers into Victoria. U.S. Army Maj. R. B. Farrow mobilized 1000 American troops and ordered extra military policemen into the bazaars to call every soldier back from leave. "My men are completely at your service," he told British authorities as convoys of troops raced for the docks.

Blast-shattered hulks of freighters etch the skyline of the Bombay waterfront in this view from Victoria Dock, where the *S.S. Fort Stikine* vanished from the face of the earth following devastating explosions.

U.S. Army Signal Corps photograph

The flood of fire engines, ambulances, troops and rescue teams collided with the torrent of dazed and powder-blackened victims streaming from Victoria and Prince's. A policeman closed Yellow Gate to keep out the curious and to stop rescue lorries from driving over hoselines laid by incoming fire apparatus. A navy chief petty officer insisted the gate remain open so ambulances could get in and out. The policeman refused. The navy won its point when the chief shoved him out of the way and posted a sentry with a rifle and bayonet to make sure the gate stayed open.

The police, convinced the blast was an act of sabotage, roped off Alexandra Dock to everyone. "Sorry," they told detachments of soldiers and sailors who drove up in lorries to start moving stores of ammunition out of the path of the flames. "You can't go in without a pass."

The burned out hull of the *S.S. Empire Indus,* which was tied up at Prince's Dock when the catastrophic explosions tore apart the ammunition ship *S.S. Fort Stikine* while it was being unloaded in wartime Bombay.

U.S. Army Signal Corps photograph

Salvage workers attempt to find bodies aboard the *S.S. General V. Sweiten,* one of the many freighters destroyed following the disaster on the Bombay waterfront.

U.S. Army Signal Corps photograph

Coombs staggered toward Yellow Gate with his few firemen who could walk and met the first of the fire engines converging on the docks. He ordered their section leaders to hook up to hydrants and start fighting fires in the godowns. Like everyone else, Coombs was too appalled by the magnitude of the explosion and the unbelievable carnage to realize that a worse catastrophe was in store for Bombay. Churchill's "Ace-in-the-Hole" was about to be trumped. Flames were gnawing deeper into the *Fort Stikine's* No. 4 hold where 1,740,000 pounds of ammunition were stowed.

At 4:33 p.m., 27 minutes after the first blast, the 870 tons of TNT in No. 4 hold blew up with an enormous roar dwarfing the first explosion. The *Fort Stikine* disappeared from the face of the earth. Farmers 50 miles inland paused at their plowing and wondered what in the name of Mohandas Gandhi was responsible for the sudden gust of hot air from the west. They knelt and prayed. The needle of the Simla seismograph wiggled crazily.

A massive tidal wave that made the first one seem like a ripple on a fish pond scooped up the 5000-ton *SS Jalapadma*, lifted her 60 feet into the air, spun her around at almost right angles to the quay, and dropped the blazing ship on top of a 4-story metal warehouse, crunching it flat. The *Jalapadma's* poop deck flew over a cluster of warehouses and whanged down into Frere Street, half a mile away. Her propeller whizzed hundreds of yards more and burrowed for half the length of its shaft into a field of kerosene drums. One hundred-ton chunks of the *Stikine* hurtled across the docks like battering rams, mashing buildings and bodies. Her

The enormous tidal wave created by the explosions of ammunition aboard the S.S. Fort Stikine lifted the 5000 ton Dutch freighter Jalapadma 60 feet into the air. It soared over a four-story warehouse before it came crashing down with its hull snapped in half.

U.S. Army Signal Corps photograph

The wrecked bow of the *S.S. Jalapadma* shown after being flung over the quay. Her back was broken, and she was a total loss.

U.S. Army Signal Corps photograph

Before the explosion, the stern of the *S.S. Jalapadma* was in the position where the bow is shown. She was moored at Pier No. 2, and the "X" at the extreme right shows the location of the *S.S. Fort Stikine*.

U.S. Army Signal Corps photograph

boiler slammed down onto trolley tracks a quarter of a mile west. The *Stikine's* 3-ton anchor zipped 1000 yards across the basin, was snared by a burning freighter's top rigging and swallowed like a fly in a spider web.

A deputy police commissioner was driving Coombs to St. George's when the steering wheel was wrenched from his hand. The auto shimmied. Coombs looked back toward the docks at the gigantic mushroom-shaped cloud of thick black smoke. Neither Coombs nor any other man alive had ever seen anything like it in that pre-atomic bomb age.

The mile-high cloud sucked up a lethal fallout of gold bars, red hot ship fragments, blazing bales of

"X" marks the spot where the *S.S. Fort Stikine* was moored at Pier No. 1, Victoria Docks. "A" shows the rubble left from the adjoining warehouse. All that remains of the *S.S. Jalapadma* can be seen in the background.

U.S. Army Signal Corps photograph

cotton, lumber, and corpses. Exploding anti-aircraft shells and tracer bullets skittered across the cloudless sky, cutting insane, zig-zag courses and leaving white vapor trails.

Coombs watched aghast as the wind drove the cloud back over Bombay. Some of the blazing debris and gold showered into the Arabian Sea but most of it fell upon the docks, kindling new fires aboard ships and among the collapsed warehouse. A carpet of burning cotton was smeared across Victoria basin.

Approximately one half mile from the explosion, this and many other residential districts caught fire from burning cotton bales, oil, and red hot pieces of steel thrown into the area.

U.S. Army Signal Corps photograph

The wind blew bits of cotton high over Coombs' head and dropped them more than a mile away, deep inside the tenement district west of the docks beyond the railway yards. Unnoticed, the smoldering puffs fluttered down on wood-shingled roofs. The wind fanned them into brisk blazes. Where there had been 10 fires after the first blast, 100 now burned.

Countless servicemen and longshoremen rushing into the docks to help with rescue work were annihilated by flying debris, bullets and shrapnel. Hundreds who had survived the first blast were slaughtered. Ten blocks from the *Stikine* Capt. Signey Kielly and a shipmate were hurrying back to their freighter when a flying saucer-like iron wedge sliced Kielly's companion in two.

A police inspector dove for shelter inside a steel pipe and was burned alive when the heat wave in the wake of the explosion turned the tube red hot. A freight train in the Great Indian Peninsular Railway Yards, 300 yards west, was knocked over.

Nurse Ketchell felt her cap fly off as she was lifted gently but firmly and pushed through an open door into the adjacent room. She landed on her feet. Dazed, she rubbed the spot on her head where the pin that had held her cap in place had snatched out a tuft of hair.

Far out in the Arabian Sea, naval officers and troops aboard the *Mariposa* rushed to the rail when their transport rocked from the blast. They looked for a periscope or an oil slick. The *Mariposa's* second officer, Edward F. Oliver, recalls that he thought his destroyer had dropped a depth charge pattern nearby.

Troops fought fires while minor explosions were still occurring.

The Times of India Press

Dense clouds of smoke billowed skyward as the blaze raged through Bombay's tinder-dry waterfront area following the blasts.

Coombs' injuries were quickly bandaged and he hurried back to the fireground. Even with his brigade at full strength it would have been impossible to stop the sea of fire spreading out in a half-moon crescent north, south and west from the waterfront, along the outer edge of the claw-shaped city's eastern pincer. Gun-shy from the sharp explosions, Coombs abandoned Victoria and the burning ships in the basins. Without fireboats he was helpless in there. He knew he could not cover all sides of the fire and decided to leave the Alexandra Dock flank on the southern tip to military firefighters. They had the manpower and trailer pumps to fight the flames and lorries to haul the ammo to safety.

At the northern tip of the crescent Coombs counted at least a dozen full-blown fires raging high into the late afternoon sky over Prince's Dock, Elphinstone Estate, and Wadi Bunder Goods Depot. The Burmah-Shell Oil Company tank farm was in the path of the fire but Coombs knew his friend, Leslie Hall, could be relied upon to take charge along that flank.

Coombs' main problem was Bombay proper, spread out in all its 94-square mile vastness beyond the westerly bulge of the flaming crescent. Along this flank he faced a life or death battle. Toward the docks the Carnac Bunder and Rice Market areas were burning furiously and sending up huge pillars of black smoke. Many of the blazing sheds and godowns in this district held foodstuffs, war materials, baled cotton, and rubber tires, all girding into a mammoth firestorm that could sweep the city.

If this husky wall of fire jumped the waterfront area and melded with the north and south tips of the crescent, it could cross Mahomed Ali Road, 1,000 yards west of the docks, and roar into the main tenement district. If this happened, nothing would stop the fire and the heart of Bombay would be burned out.

Coombs placed his depleted manpower, all the volunteers he could muster, and equipment in a thin arc around the western perimeter of the fire and prepared to attack head on. He knew it was pointless to hope that fires started by flaming debris pitched into the tenement and market districts bordering the docks would be doused by native bucket brigades. Thousands of Bombayans had fled in panic westward toward the center of the city, ignoring in their haste the bars of gold littering the cobblestoned streets.

As the sun disappeared behind Malabar Hill on Bombay's western pincer, and night settled over the city, Coombs saw an ominous glow rising from deep inside an area of closely built tenements to the east of Mahomed Ali Road. The smoldering cotton puffs were taking root.

Coombs' sense of utter helplessness increased tenfold when one of his pumpers hooked up to a hydrant in the warehouse district. Only a trickle came from the nozzle. A 22-foot-long spear of iron from the disintegrating *Stikine* had imbedded itself in Frere Street, severing a main trunk water line supplying hydrants in the area.

"Damn the luck!" cursed Coombs. "We'll have to pump from the basins and relay water to the front lines." When the firemen dropped their suctions off the docks the strainers quickly became clogged with floating cotton. They kept their pumpers from choking by skimming away the cotton with brooms.

Wherever Coombs and his men found water and drove their streams into the fire the flames contemptuously mauled them backwards. Coombs conceded one block of buildings after another while he set up what looked to be a sturdy line of defense further west along Argyle Road. The fire bulled past that line, too. Millions of sparks soared high into the darkness over the crown of flames and were driven far behind Coombs' lines to build new fires.

Drivers of rescue lorries and ambulances braved the heat and dense smoke, searching in the fire's glow for victims. Corpses were stacked like sacks of potatoes in the morgue. When the pile reached the ceiling, city rubbish trucks hauled the bodies away for mass cremations by committees hurriedly formed by Hindu and Moslem leaders.

It was now 6:30 p.m., in Bombay, two hours after the explosions. The city's skyline, etched by flame, was visible for more than 100 miles. A vast blanket of smoke swirled over Bombay. The wind blew it far inland where farmers smelled the burning city and gazed reverently at the ominous glow piercing the night off to the west.

Coombs damned the low water pressure and the workers for taking so long to build an emergency by-pass around the break. He realized, though, that even with the hydrant system in full operation the weak pumping capacities of his equipment could not possibly throw the massive streams of water that his ruined heavy duty engines could. He needed a lot more men and pumpers—any kind of pumpers—to effectively surround the fire.

In the heat of battle neither Coombs nor anyone else remembered the hundreds of privately owned fire engines standing idle in factories and textile mills throughout Bombay. Their firemen were listening impatiently for the city's radio station to explain the two explosions they had heard three hours earlier and the smoke and glow now lighting the sky toward the docks.

Night view of Bombay burning looking towards the Customs House and Victoria and Prince's Dock areas.

U.S. Army Signal Corps photograph

So long as nobody called them, these firemen figured Coombs had the situation well in hand. They went to bed.

Shortly after the first blast, the government had clamped a lid of secrecy on all information media.

The city's radio could have asked this additional manpower to report with their rigs to specific hot spots along the waterfront where they might have saved rows of buildings otherwise doomed to the flames. Shortly after the first blast, however, the government had clamped a lid of secrecy on all information media. The city's radio station continued to broadcast ballet music, symphonies and items of news from everywhere except Bombay. Newspaper and wire service reports were held up by the government censors. The enemy would learn soon enough of this mortal blow to the Allied supply line and there was no point in alarming our troops on the fighting fronts.

The government could keep the news from the world but it could not stop rumors from traveling enormous distances. By 7:30 that night sailors aboard ships of the British Royal Navy cruising in the Persian Gulf heard that Bombay was obliterated by an air raid. Not long after, Tokyo Rose told her listeners that Bombay lay in smoldering ruins, thanks to the clever work of skilled Japanese saboteurs. She compared "this brilliantly executed attack" with Pearl Harbor.

Nightfall in Bombay brought the usual mysteries of darkness. The fiery cloud over the city, coupled with the blaring silence of the official radio, spawned an epidemic of rumors, many of them encouraged by foreign agents known to be in the city. The rumors took on all shades of meanings when more and more natives learned the telephone system in the docks was not working, that electrical sub-stations were gutted and that the water pumping station was in wrecked darkness. The populace wondered if saboteurs had also tampered with the air raid sirens so they couldn't warn Bombay of the attack.

To head off a threatened stampede and to allay worried diplomatic inquiries from Allied capitals—especially No. 10 Downing Street—Bombay's governor, Sir John Colville, finally issued an announcement. Even granting the stoic calm of the British in times of turmoil, Sir John's remarks must go down as the classic understatement of World War II:

"About 4 o'clock this afternoon an accidental fire broke out in a ship in the docks and before it could be brought under control it spread to some ammunition and caused violent explosions. The fire also spread to adjacent godowns. The fire services, the military and ARP Services were on the scene immediately and as a result of their efforts the situation is now under control. A number of casualties have unfortunately occurred and the injured are being attended to in hospitals."

The announcement was printed in its entirety in the London *Times* at the bottom of page 4. Three days later the New York *Times* which prints "All The News That's Fit to Print," carried a four-paragraph United Press item on page 3.

At 8 p.m., seven hours after the fire was discovered and four since the first explosion, Sir John called an emergency meeting of government officials at Bombay Police Headquarters. Coombs hurried to the meeting after stopping off at St. George's to have his soot-streaked bandages changed.

"I realize all of you have done the best you can," said Sir John, "but the fire continues to burn closer to the center of Bombay. I propose we use dynamite to blast a firebreak before the city goes up in smoke. Let's try a quarter-mile wide strip down in the tenement district."

"Begging your leave, Sir John," said Coombs, "dynamite won't do the trick. It would start more fires than it could ever stop." Coombs explained that the blasts would shatter windows in the area adjacent to the firebreak and offer open invitations to firebrands.

"Before we did any dynamiting down there we'd have to evacuate those tenements. Thousands of blokes haven't gotten out of there yet. The only way you'll get them out is by force or at bayonet point and I don't fancy that idea sitting well with the Home Office. The natives won't stand by while we blow up their homes and a riot is something I'd hate to worry about at this stage!"

"So if we don't dynamite. What then?"

"With luck and if that wind ever changes we should be able to hold on the western bulge of the fire, the flank that's working toward Mahomed Ali Road."

The military, government, police, Port Trust and other officials present agreed that Coombs' strategy was the only possible one. Sir John declared the meeting closed.

Down in the basins, meanwhile, volunteer naval parties were saving burning ships with small hoselines operating off the vessels' own pumps and portable rigs put aboard damaged but navigable tugs. British Royal Navy Lt. W. H. Ellames took over an abandoned towboat and, sweeping Victoria basin with the boat's searchlight, spotted the bobbing heads of 32 men whom he rescued from the cotton-covered water. Sub-Lt. C. A. Stocken saved 60 men by maneuvering his launch in and out of the docks.

The smoke was too thick for Coombs to have known this from his vantage point and those fighting to save ships and lives in the basins were too busy to send messengers back for help. Consequently, ships and men who might have been saved by additional firefighting parties were left to the ravages of the flames. Many drowned in the murky waters.

In Alexandra a handful of tars from the *HMS Sussex*, together with harbor officials and soldiers, gingerly stepped over charred corpses as they boarded 16 freighters, some of them burning, including the three ammo ships. The work parties put out the fires and with tugs nudged the ships out of the harbor. Other volunteers boarded the oil tanker and found her thick deck plates had nearly been punctured by a flying piece of iron. With Dockmaster A. N. Thompson's help they maneuvered her out to open sea where they saw in the distance the blazing silhouettes of a string of cotton lighters that had drifted from Victoria basin. When the tide turned the boats wandered back to shore.

By midnight Coombs was battling the flames building by building, street by street. His back throbbed with pain. During his years as a fireman he had never been whipped by a fire, but he could not so much as second-guess the twists and turns this one was taking. Army tanks bulled openings in debris-clogged streets so Coombs could move his fire engines and men forward to meet approaching walls of flames.

The broken main in Frere was finally bypassed. "Raise the pump pressure all you can on these lines!" Coombs shouted over the bellowing flames. Despite the increased pressure, the streams still turned to steam before they bored into the heart of the conflagration. Heavy clouds of smoke and waves of heat pushed the firemen back.

A brigade section leader ran up to Coombs and handed him a flame-blacked slab of metal, an inch and a half thick, about 3 inches wide and 15 inches long. "By George, you're right!" said Coombs, scraping away the soot. "It's a hunk of solid gold!" Turning the bar over he saw that the explosion had driven packing crate nails deep into the bar.

In Alexandra Dock, where the police had finally admitted working parties, the job of removing ammunition from five sheds was proceeding at a desperate pace. The city's racial tensions had ceased to exist so far as a thousand white and Negro servicemen working side by side were concerned. Troops handed 155-pound crates of ammo across human chains and into waiting trucks. When loaded, the convoys raced under motorcycle escort from Bombay. Deputy Docks Director L. E. Walsh found a trainload of ammunition standing beside a burning godown. He commandeered a locomotive, coupled it, and the train highballed out of the city. Before dawn more than 22,000 tons of TNT were removed from Alexandra.

Meanwhile the fire continued to shove Coombs and his men back. The flames were working perilously close to Mahomed Ali Road. He was directing his men in placing hoselines in the lower floors of a 6-story tenement where a small fire had been started by flying sparks when, before he knew it, the flames raced up the flimsy partitions to the attic where they crowned, burst through the roof and spilled fire onto adjacent tenements. In five minutes the whole street was blazing.

"Fall back! Fall back!" yelled Coombs. "Drop your lines and run. Abandon those rigs!"

Coombs made a determined stand in Dongri Street by massing a dozen hoselines and forming a water curtain, but this line, too, crumbled in the face of the rampaging firestorm. Brick walls swayed ominously while Coombs and his men quickly dragged hoselines and pushed fire engines back from danger. Seconds later the walls would come thundering down, catapulting burning debris into other buildings across the narrow alleys and streets. Immediately, new rows of structures spouted flames.

Small fires consolidated into bigger blazes and leaped into adjoining blocks of tenements, stores and warehouses before amalgamating with other advancing wings of the fire. The Akali Dal, a volunteer organization of Sikhs, armed themselves with clubs to prevent looting of stores, only to surrender the buildings to the greedy flames. A few blocks miraculously escaped and stood out as gaunt islands in the smoldering rubble. But their reprieve wouldn't last long.

At 3 o'clock in the morning, 13 hours after the fire was discovered, Coombs watched the fierce smoke boiling from a four-story brick godown suddenly blossom into brilliant lavender blue flames as the fire reached stores of Navy grog. From off to his left an overpowering stench was blown over the firelines as flames touched off a warehouse loaded with ghee, a clarified native butter. In another godown the fire exploded hundreds of acetylene gas cylinders which skyrocketed, slashing flaming wakes in the dark sky high above the fire cloud.

Firemen clanging through the smoke-filled rubble-littered streets to sectors where help was badly needed encountered other major fires en route. Nobody was fighting these outbreaks so they stopped and set up firelines, leaving the other firemen to do the best they could. This sort of distribution of firefighting forces had happened during the London Blitz but Coombs never expected to find himself facing the same problem.

Shortly before dawn Coombs suddenly noticed the smoke was no longer being driven so hard into his face. The soaring sheets of flame which had been slanting toward him now stood straighter. In places, they were bending back toward the docks.

"The wind's changed," breathed Coombs, gratefully. "We've got a fighting chance to save the city."

Coombs would not permit himself any real optimism, for at that instant another pitch dark street of four-story brick warehouses exploded into flames. Again he ordered his men to retreat. At best, Coombs could only hope to surround the crescent-shaped perimeter of the fire and continue to fight a delaying action. The fire would gradually burn itself out if the wind did not change again.

Coombs hurried over to Prince's Dock and marveled at the brilliant stop made by Hall and his small group of volunteers using a few low-pressure trailer pumps which they had dragged over 12 railroad tracks. The situation was also looking better in Alexandra at the south tip of the crescent.

Coombs was barely aware of the dawn. A thick pall of smoke hovering over Bombay nearly eclipsed the rising sun. The flames were beginning to exhaust themselves and fall back to feed upon buildings missed during the night. Fingers of fire probed Coombs' lines for soft spots before working back into the burned-over area. Rubble made the streets too impassable and the danger of falling walls was too great for him to risk his men and equipment in pursuing the fire.

By 10:30 Saturday morning, 20 hours after Mahomed Taqi discovered wisps of smoke among some cotton bales, the worst was over, and wetting-down operations started. Coombs decided to get some rest. He knew Bombay would not be out of danger for days. The Ballard Estate district, the city's main business area, would not be safe for 48 hours. The flames in the warehouse, tenement and dock areas would not be fully controlled for two weeks, and fires in the grain and rice market district would smolder for months.

Sixty-six of the brigade's 156 officers and men were dead or dying and 79 badly hurt.

Bombay was saved. The fire had not crossed Mahomed Ali Road. Coombs had won, but at a terrific cost. Sixty-six of the brigade's 156 officers and men were dead or dying and 79 badly hurt. Only 11 had escaped injury and most of these had worked in the control room deploying manpower and apparatus. The explosions, fires and toppling walls had ruined 35 of the brigade's 72 pieces of firefighting equipment. Coombs gazed across the smoldering 100-acre wilderness where 586 buildings had stood less than 24 hours earlier. On his way home he drove by the docks. Nineteen of the 39 ships in port when

Charred bales of cotton float among ships that were destroyed following the explosion aboard the *S.S. Fort Stikine*, which was loaded with ammunition, cotton, and gold bars.

the fire started were still burning. A gray haze hung over the basins and a carpet of cotton bales, chunks of cargo, debris, cinders and charred corpses bobbed on the surface. Coombs saw no trace of the *Fort Stikine*. All that remained of Berth I was a vast water-filled crater bordered by a pile of broken cement where the quay used to be.

Bombay counted its dead and injured. The mayor was released from jail to help set up a welfare committee to assist the Red Cross with the 50,000 homeless. Government estimates placed the loss of life in excess of 800 people and 2400 hurt, adding that these figures were "not an overstatement." An official United States Coast Guard report states that between 1000 and 1500 people were killed and 4000 injured. It estimated damage in excess of $1 billion. Many who were in Bombay at the time say the Coast Guard figures are more accurate.

There is no such thing as an accurate census in Bombay. No doubt hundreds of anonymous persons were blown to bits or incinerated. Corpses were still being found months after mopping up operations started.

The blow to the war effort was staggering. More than 100,000 tons of vital shipping were put out of action. Eleven freighters were a total loss, and nine others badly damaged. Small wonder the full details of the catastrophe were not immediately divulged.

We can only speculate at what the cost might have been or what effect the disaster might have had upon the Turkish front or even upon subsequent Russian domination of Eastern Europe. Bombay would have been an important supply link in the planned invasion of Japan. Certainly in April, 1944, nobody knew the war would come to an abrupt end 16 months later after two other historic explosions.

Reconstruction started immediately. More than 10,000 workers, most of them British, Indian, American, and West African soldiers, cleared 900,000 tons of debris from the area in the next 6 1/2 months. A few gold bars were found in the junk.

When the docks were reopened, a Royal Commission of Inquiry issued two reports on the catastrophe after questioning 133 witnesses and examining a large number of documents at secret hearings. The commission's reports cited five major reasons for the disaster:

The propeller and tail shaft of *S.S. Fort Stikine* after being lifted from the dock bottom where it was blown by the second explosion.

1. The existence of a state of war resulting in the practice of bringing ships loaded with explosives and ammunitions into the docks.

2. When Royal Air Force planes were unloaded at Karachi the available space aboard the *Stikine* was used for high priority cotton consigned to Bombay mills. The cotton was stowed haphazardly above and below the ammunition. (American shipping practice forbids the carrying of cotton and explosives on the same ship).

3. "Accidental ignition of cotton." The commission investigated the possibility of sabotage. Ruling it out, they concluded the fire was probably started by a careless workman smoking during the lunch hour.

4. The commission sharply criticized the failure to size up the seriousness of the problem at the start and to "take energetic steps either to extinguish the fire or to take alternative action."

5. The board also criticized the absence of a "centralized executive control at the scene with power to issue paramount orders and coordinate the various authorities concerned."

The government noted Coombs' "praiseworthy heroism" and for his valiant role he was awarded a bar to his police and Fire Service Medal. Coombs later retired.

Neither of the commission's reports mentioned the $5,600,000 worth of gold aboard the *Stikine*. Even today the Reserve Bank of India would rather not talk about it, either. Inquiries to the Indian Government go unanswered.

Nobody will say how many bars have been recovered. At least three were returned soon after the explosions: the one that landed at Motliwala's feet, the one found by Coombs' men and a third which mysteriously turned up in Trichinopoly, 1000 miles southeast of Bombay.

Although half a century has passed since the disaster, gold ingots continue to be found, sometimes by dredges scraping the muck of the basins. A standing reward of 999 rupees, about $210, goes to each finder.

Reliable estimates along Bombay's waterfront say more than 90 bars are still missing. If you're willing to settle for the reward money, you stand to collect more than $20,000. If not, there's better than $3,690,000 worth of hidden treasure scattered along Bombay's waterfront and in her rubbish dumps awaiting some lucky finder.

AFTERWORD

Since the dawn of civilization, fire has been the most potent war weapon, especially in cities where firestorms can destroy war production and shipping while dissolving any ability or desire to continue. Hitler never learned this truism. Nor did Imperial Japan until too late. The surprise attack upon Pearl Harbor, devastating as it was, only kindled America's resolve to win the war against totalitarianism.

It is popularly held that the atom bombs dropped upon Hiroshima and Nagasaki ended the war. They only hastened it. As the U.S. Bombing Survey teams brought out: Japan was beaten by incendiary raids on Tokyo and 64 other cities. More Japanese civilians were killed on the homefront than total casualties on all battlefronts.

The Survey further noted that air raids on German cities left citizens eager to end the war. Hitler refused to see and learn from his own people.

In the United States, meanwhile, officials worried more about saboteurs than they did over the carelessness, negligence and sheer stupidity that resulted in the deaths of hundreds of Americans.

The United States still has not learned the dangers of fire. We talk a good fight, but we do little to make it happen. Our annual per capita fire losses continue to exceed those of any other nation.

The American fire establishment has never presented a unified front against unacceptable fire safety risks. Too frequently it is torn by petty dissensions and jealousies while lobbyists at all levels of government thwart common sense solutions to fire problems facing every community.

Fire chiefs largely remain almost inarticulate when it comes to convincing their constituents—the very people who pay their salaries—and politicians who suffer such abominations as insufficient manning of apparatus, tinder-dry wood shake and shingle roofs, plus the need for built-in fire safety in all structures, especially self-closing doors, automatic fire sprinkler systems and alarms. When the fire establishment does speak out it is usually a matter of singing to a choir which attends the same meetings and reads the same specialized magazines they do.

As evidence that it can be done, we need look no further than the public acceptance and use of smoke detectors. If we are going to die in a fire it will most likely occur in our homes. Smoke detectors have saved countless lives.

Firefighting during World War II resulted in many innovations. London, in particular, demonstrated the value of pre-fire planning prior to the war. It did not solve all problems during the London Blitz, but it certainly mitigated them when as many as 3000 fires were burning at one time.

Germany and Japan paid a terrible price for the smugly traditional business-as-usual administration and operation of fire departments. No wonder fire won the war in Japan where hand-carried, hand-operated fire engines were the primary lines of defense.

Looking back some 50 years, the innovations of the London Fire Brigade during the Blitz are commonplace in American cities, although many took a long time before they got around to adopting them: Backup water supplies and incident command posts where officers devise firefighting strategies at major fires. Mobile communications vans and rear-mounted aerial ladders eventually proved their value in New York and other cities. Despite their scarcity during wartime, radio communications systems today are more frequently used than large caliber hose streams.

Japan—which suffered six conflagrations during the years prior to the war—learned many lessons. Today's visitors to Tokyo marvel that the city's fire department is among the world's best. Many facets of Tokyo's fire protection, public information and public attitudes toward fire could well be emulated in the United States.

Our global economy must discard traditional parochialism which claims: "If my fire department doesn't think of it first then it must be no good."

Paul Ditzel
Woodland Hills, California

INDEX

Note: Italicized numbers refer to photographs

INDEX

ABOUT THE AUTHOR

PAUL DITZEL

Paul Ditzel, widely acclaimed as the "dean" of fire service authors, has written 15 books and over 600 articles for magazines, including the *Reader's Digest*. He was a contributing editor to *Firehouse Magazine*, and *Fire Engineering*. His best-selling book, *Fire Engines, Firefighters,* was submitted for a Pulitzer Prize in American History.

He holds a Master of Science Degree from Northwestern University, where he graduated with highest scholastic honors. In 1967 he was appointed as the Los Angeles Fire Department's first civilian fire inspector.

Paul Ditzel brings to the Fire Buff House series of fire service history books the very finest research and colorful writing on firematic subjects. Additional volumes are in preparation.

Other Books in the Fire Service History Series from Fire Buff House Publishers.

- ***CHEMICAL FIRE ENGINES*** by W. Fred Conway
  The only book ever written about these amazing engines that for half a century put out 80% of all fires in spite of the fact that they never did perform as advertised! Hard cover. 128 pages, over one hundred photographs and drawings.
- ***FIREBOATS*** by Paul Ditzel
  The definitive work on the history of marine firefighting - a book so real you can almost hear and feel the engines and pumps throbbing! Relive the thrilling accounts of fireboats in action at spectacular fires. Hard cover. Over a hundred pictures.
- ***FIRE ALARM!*** by Paul Ditzel
  The story of fire alarm telegraphy - a nostalgic overview of fire alarm transmission in America. Includes the incredible story of an obscure telegraph operator, John N. Gamewell, who built a vast fire alarm empire with a 90% market share. Hard cover. Hundreds of pictures.
- ***LOS ANGELES FIRE DEPARTMENT*** by Paul Ditzel
  The fascinating complete history of the Los Angeles Fire Department. America's foremost fire historian details the major fires during the city's history, including the Bel-Air brush fire disaster, and the Central Library blaze. Oversize book with 250 pages and hundreds of photographs of fires and apparatus. Soft cover.
- ***A FIRE CHIEF REMEMBERS***
  by Battalion Chief Edwin F. Schneider (Ret.)
  The poignant, touching, moving account of his FDNY career. Now 82 years of age, Chief Schneider looks back at the series of incredible events in his 34 year career climb from buff to chief. Soft cover. Illustrated.
- ***LAGUARDIA'S FIRE CHIEF*** by Kathleen Walsh Packard
  The fascinating biography of New York City Fire Chief/Fire Commissioner Patrick Walsh, who served under the mercurial mayor, Fiorello LaGuardia. Chief Walsh and Mayor LaGuardia would race to see who could get to the fire first. This account of Chief Walsh's storybook career is written by his granddaughter. Soft cover. Illustrated.
- ***DISCOVERING AMERICA'S FIRE MUSEUMS***
  by W. Fred Conway
  A comprehensive guidebook describing 150 museums in the United States and Canada displaying vintage fire apparatus, equipment, and memorabilia. Includes a history of firefighting apparatus and equipment from Colonial times to the present. 188 pages, 200 illustrations, soft cover.
- ***FIREFIGHTING LORE***
  by W. Fred Conway
  Probably the most unusual book of firefighting history ever written. It recounts dozens of amazing, strange, and incredible stories from firefighting's past and present — every one true. 160 pages, illustrated, soft cover.

These books are available wherever fire books are sold,
or you may order direct from the publisher.

Fire Buff House Publishers
P.O. Box 711, New Albany, IN 47151-0711

1-800-234-1804